ENERGY POLICIES, POLITICS AND PRICES

CRUDE OIL PROPERTIES AND THE HAZARDS OF TRANSPORT

BACKGROUND, DATA AND LITERATURE SUMMARY

ENERGY POLICIES, POLITICS AND PRICES

Additional books in this series can be found on Nova's website under the Series tab.

Additional e-books in this series can be found on Nova's website under the e-book tab.

ENERGY POLICIES, POLITICS AND PRICES

CRUDE OIL PROPERTIES AND THE HAZARDS OF TRANSPORT

BACKGROUND, DATA AND LITERATURE SUMMARY

DIANNE GIBBS
EDITOR

nova publishers
New York

Library of Congress Cataloging-in-Publication Data

ISBN: 978-1-63483-070-6

Published by Nova Science Publishers, Inc. † New York

CONTENTS

PREFACE

Several fiery rail accidents in 2013-2015 in the U.S. and Canada carrying crude oil produced from the Bakken region of North Dakota have raised questions at many levels on the safety of transporting this, and other types of crude oil, by rail. Sandia National Laboratories was commissioned by the U.S. Department of Energy to investigate the material properties of crude oils, and in particular the so-called "tight oils" like Bakken that comprise the majority of crude oil rail shipments in the U.S. at the current time. The book provides a literature survey of public sources of information on crude oil properties that have some bearing on the likelihood or severity of combustion events that may occur around spills associated with rail transport. The book also contains background information including a review of the notional "tight oil" field operating environment, as well a basic description of crude oils and potential combustion events in rail transport.

In: Crude Oil
Editor: Dianne Gibbs

ISBN: 978-1-63483-070-6

Chapter 1

LITERATURE SURVEY OF CRUDE OIL PROPERTIES RELEVANT TO HANDLING AND FIRE SAFETY IN TRANSPORT*

David Lord, Anay Luketa, Chad Wocken, Steve Schlasner, Ted Aulich, Ray Allen and David Rudeen

ABSTRACT

Several fiery rail accidents in 2013-2015 in the U.S. and Canada carrying crude oil produced from the Bakken region of North Dakota have raised questions at many levels on the safety of transporting this, and other types of crude oil, by rail. Sandia National Laboratories was commissioned by the U.S. Department of Energy to investigate the material properties of crude oils, and in particular the so-called "tight oils" like Bakken that comprise the majority of crude oil rail shipments in the U.S. at the current time. The current report is a literature survey of public sources of information on crude oil properties that have some bearing on the likelihood or severity of combustion events that may occur around spills associated with rail transport. The report also contains background information including a review of the notional "tight oil"

* This is an edited, reformatted and augmented version of a report, SAND2015-1823, issued by Sandia National Laboratories, March 2015.

field operating environment, as well a basic description of crude oils and potential combustion events in rail transport.

NOMENCLATURE

AFPM	American Fuel & Petrochemical Manufacturers
API	American Petroleum Institute
ASTM	American Society for Testing and Materials
bbl/d	Barrels per day
BLEVE	Boiling liquid expanding vapor explosion
BPP	Bubblepoint pressure
CCE	Constant composition expansion
CCQTA	Canadian Crude Quality Technical Association
CFR	Code of Federal Regulations
DOE	Department of Energy
DOT	Department of Transportation
DVPE	Dry vapor pressure equivalent
EIA	Energy Information Administration
EOS	Equation of state
EPA	Environmental Protection Agency
FE	Office of Fossil Energy
FPC	Floating piston cylinder
FRA	Federal Railroad Administration
GOR	Gas-oil ratio
GPA	Gas Processors' Association
HMR	Hazardous Materials Regulations
HT	Heater treater
IBP	Initial boiling point
ISO	International Organization for Standardization
LFL	Lower flammability limit
LPG	Liquefied petroleum gas
MSDS	Material Safety Data Sheets
NDIC	North Dakota Industrial Commission
NDPC	North Dakota Petroleum Council
PG	Packing group
PHMSA	Pipeline and Hazardous Materials Safety Administration
PRD	Pressure relief device
psi	Pounds per square inch

psig	Pounds per square inch gauge
RVP	Reid Vapor Pressure
SDS	Safety Data Sheet
SEP	Surface emissive power
SNL	Sandia National Laboratories
SPR	Strategic Petroleum Reserve
TDGR	Transportation of Dangerous Goods Regulations
TVP	True Vapor Pressure
U.S.	United States
UFL	Upper flammability limit
V/L	Vapor-liquid
VLE	Vapor-liquid equilibrium

1. Executive Summary

The U.S. Department of Energy commissioned a technical team[1] through Sandia National Laboratories to examine the properties of tight crude oils as they relate to potential combustion events in the rail transport environment. Key objectives of this investigation are to characterize and define tight crude oils based on their chemical and physical properties, and identify properties that could contribute to increased potential for accidental combustion. This investigation was commissioned in response to the occurrence of several rail accidents involving crude oil combustion in the U.S. and Canada during 2013-2014, some of which involved loss of life, property damage, and environmental impacts. The repeat appearance of crude oil as a factor in rail accidents is a relatively new phenomenon, as U.S Department of Transportation data indicate that more oil was spilled from rail cars in the single year 2013 than the total spilled from railcars in the 30 years prior. A significant factor behind this increase in spilled oil is the increase in oil-by-rail transit in the U.S., which the American Association of Railroads reported as 9,500 carloads in 2008, growing to 407,761 carloads in 2013. This recent increase in oil-byrail volume is in response to the dramatic increase in U.S. production of crude oil from "tight" shale formations, with rail helping to close gaps in transport capacity where pipelines and trucks cannot meet demand.

[1] Team includes Sandia National Laboratories, University of North Dakota Energy & Environmental Research Center, Allen Energy Services, and GRAM, Inc.

As the first step in the investigative process, this initial report is a compilation and summary of publicly available literature and data pertaining to the chemical and physical properties of tight crude oils. Key literature/data sources reviewed include recent reports on Bakken crude properties commissioned by the American Fuel & Petrochemical Manufacturers, North Dakota Petroleum Council, and U.S. Department of Transportation Pipeline and Hazardous Materials Safety Administration, and data from the U.S. Strategic Petroleum Reserve. This initial investigation identified gaps in important crude oil characterization data, uncertainty regarding how best to sample and analyze crude oil to ensure that its properties are accurately determined, and deficiencies in the understanding of how crude oil properties impact its potential for accidental ignition, combustion, and explosion.

Based on these findings, it is recommended that a subsequent effort focus on preparation of a report that comprises:

- A sampling and analysis plan to close gaps in existing knowledge of tight crude oil properties.
- An experimental plan for numerical modeling of combustion events and testing from small to large scale to obtain data for hazard evaluation.

Key findings of this initial literature-based investigation are summarized below.

1. Due to significant variability in criteria and procedures utilized in selection, acquisition, and analysis of crude oil samples, the available data are of insufficient quality to enable a meaningful comparison of crude oils—either to each other or against a designated standard.

Although a large volume of crude oil characterization data is available for review, the samples used to derive the data were acquired from a wide range of supply chain points using a variety of open and closed sampling techniques and characterized via a variety of analysis methods. In many cases, limited information was provided regarding where and how samples were acquired and analyzed. In addition to making oil-to-oil comparison difficult (especially for tight oils that typically contain significant quantities of dissolved gases and volatile liquid species), this variability makes it difficult to establish a set of properties that could be used to define a certain oil type and/or identify an oil as—for example—a conventional oil, or a tight oil.

2. In addition to variability due to sampling and analysis methods, variability may also be introduced through crude oil conditioning, storage, and transport.

"Conditioning" refers to processing conducted—typically at or near the well site—to remove crude oil impurities prior to transport. Impurities include gases, water, and solids that were produced with the crude oil. Although a fairly limited range of equipment is utilized to effect removal of gases from crude oils, the equipment can be operated at varying levels of rigor to meet specifications for transport and sale. Variations in oil reservoir conditions, well-site processing operations, residence time in storage tanks, and transportation method (truck, pipeline, or rail) may all influence crude oil composition and properties.

3. Currently used methods for assignment of crude oil transportation hazard classification and packing group are often inadequate.

As mandated by current federal law, hazard classification and packing group assignment is done on the basis of crude oil initial boiling point (IBP) and flash point; however, there was widespread agreement among the studies reviewed that the methods commonly used for IBP determination are inappropriate for application to crude oils, especially tight crude oils containing significant quantities of dissolved gases. Major reasons for inappropriateness include 1) the use of liquid condensation temperature to directly establish IBP (problematic with crude oils that contain noncondensable gases) and 2) the use of gas chromatography (GC) to indirectly establish IBP (problematic because GC effects clean separation of individual crude oil gas and liquid species and does not account for the very real impact of azeotropes and other intramolecular effects on IBP). While recommendations for improved methods have been offered, no widespread agreement has been reached regarding the adoption of more appropriate methods.

4. Relationships between crude oil properties and probability or severity of combustion events in rail car spill scenarios have not been established.

Although it is likely that a combination of crude oil properties—especially those associated with potential for flammable vapor formation—could be used to predict combustibility, no specific, objective data were found that correlated known crude oil properties with the likelihood or severity of rail transport-related combustion events. While industry groups actively working on this problem have been identified, their progress and results have not yet been released to the public.

5. General lack of uniformity in methods and QA/QC across industry makes comparison of crude oil vapor pressure difficult.

There is some literature consensus that vapor pressure of a "flammable liquid" is a property of interest, as the vapor phase evolved from a liquid actually burns rather than the liquid phase. Several studies have put an emphasis on measuring the vapor pressure of Bakken oils across a range of locations in the supply chain. Although a large volume of vapor pressure data have been collected and reported, the wide variability in sample collection methods (open/unpressurized versus closed/pressurized) and measurement techniques (Reid versus VPCRx) utilized make meaningful comparison difficult, because differences seen among oils in a given study or across studies may be attributable to sampling and measurement techniques rather than base properties of the oils themselves. Recognition of this problem is growing, and standards do exist for sample collection and analysis; however, an appropriate protocol is still being debated via research publications and industry conferences.

6. Bakken crude is a light, sweet oil that exhibits a statistically higher true vapor pressure than the slightly heavier, blended sweet and sour oils that are stored at the U.S. Strategic Petroleum Reserve (SPR).

The SPR inventory was used as a basis for comparison with Bakken due to (i) the availability of SPR data, and (ii) the fact that current SPR inventory represents a blend of medium-light (API gravity 30-38) oils that were accumulated over the last ~20 years of SPR operation from many sources. The body of Bakken samples from public literature was found to exhibit a statistically higher mean true vapor pressure than the SPR oils, likely due to its slightly higher mole fractions of ethane, propane, and butanes. True vapor pressure data for other comparison oils (conventional production, benchmarks crudes, other tight oils) were not readily available, and should be pursued in order to build better context for how Bakken and other tight oils compare with other crude oils and flammable liquids currently in the supply chain.

7. Numerous combustion events can occur from an accident involving hydrocarbons and hydrocarbon mixtures including crude oils, with severity dependent on the amount of fuel involved, surrounding infrastructure, and environment.

Possible combustion events include:

- Pool fire, which results from the burning of a liquid fuel pool.
- Boiling liquid expanding vapor explosion (BLEVE), an explosion resulting from the failure of a vessel containing a liquid at a temperature significantly above its boiling point at normal atmospheric pressure.
- Fireball, which refers to partially pre-mixed diffusion flames that rapidly combust due to enhanced turbulent mixing and atomization.
- Deflagration: Classification of an explosion. Burning of a fuel-air mixture where the flame travels at subsonic velocities.
- Detonation: Classification of an explosion. Burning of a fuel-air mixture where the flame travels at supersonic velocities.
- Flash fire, which refers to the burning of a fuel vapor cloud that was ignited at a location away from its release point.
- Flare, which refers to the burning of fuel vapors at the source of a release.

8. No single parameter defines the degree of flammability of a fuel; rather, multiple parameters are relevant.

While a fuel with a lower flashpoint, wider range of flammability limits, lower auto-ignition temperature, lower minimum ignition energy, and higher maximum burning velocity is generally considered more flammable, the energy generated from an accident has the potential to greatly exceed the flammability impact of these and any other crude oil property-based criteria.

2. Problem Statement

Sandia was tasked by the United States (U.S.) Department of Energy (DOE)-Office of Fossil Energy (FE) to investigate the properties of crude oils that may have a bearing on the combustion potential in handling and transport scenarios. Public interest in this area has grown over the last few years largely due to the increase in awareness brought by several fiery rail accidents that involved crude oil, particularly crude oil that originated from the tight oils formations such as the Bakken region of North Dakota. A number of investigations have been launched by government (PHMSA, 2014) and industry groups (Auers, et al., 2014; Wybenga, 2014) looking at safety issues around rail transport of crude. The general conclusions from those reports vary on several key issues: (i) whether Bakken oils are more volatile than other crudes produced or shipped in the U.S., (ii) whether Bakken oils are, in turn, more ignitable or flammable, and (iii) whether Bakken and/or tight oils in general pose a greater degree of hazard in transport that would require additional testing or mitigation steps. The reports do agree that the Bakken is properly classified as a Class 3 flammable liquid under the current classification system, though there is general concern that the current means to determine packing group for flammable liquids through standard initial boiling point and flashpoint measurements give highly variable results and may not be the most appropriate methods when applied to oils with light components (ANSI/API, 2014; Murray, 2014).

The purpose of this study is to evaluate the current state of understanding of crude oil properties in the transport system and determine how these relate to hazards associated with handling and transport, as well as the associated combustion potential. The overall problem scope may be divided into three phases, listed below:

1) Problem definition phase (currently authorized work scope)

a) Define crude oil properties that have a bearing on handling and transport safety with attention to flammability risks in spill scenarios

2) Experimental phase (possible future work scope)

a) Measure parameter ranges for relevant crude properties in transport system, compare with literature and other parallel efforts (Pipeline and Hazardous Materials Safety Administration (PHMSA), American Petroleum Institute (API), Canadian Crude Oil Quality Technical Association (CCQTA))

b) Explore if/how these properties affect the degree of hazard realized in scenarios where fire may be involved

3) Application phase (all stakeholders)

a) Utilize knowledge gained during above phases to inform decisions on industry best practices, standards, and regulatory requirements in order to assure safe, economical transport of crude to market

The currently authorized work scope under the problem definition phase includes three tasks: (i) Literature Survey, (ii) Experimental and Sampling plan, and (iii) Outreach and Peer Review.

2.1. Scope of Report

The current report is the literature survey identified under the problem definition phase 1 of this project. This report comprises a review of publicly available studies on crude oil properties and fire science work relevant to tight crude oils and the operating environment relevant to transport from the producing field into the rail system.

3. Operating Environment Relevant to Study

3.1. Definition of Tight Oils

The U.S. Energy Information Administration (EIA) defines "tight oil" as "oil produced from petroleum-bearing formations with low permeability such as the Eagle Ford, the Bakken, and other formations that must be hydraulically fractured to produce oil at commercial rates; shale oil is a subset of tight oil" (2014a). The term "light tight oil" is also used because the oil produced from these formations is light crude oil. Light crude oil refers to low-density petroleum that flows freely at room temperature. Light oils have a higher proportion of light hydrocarbon fractions resulting in higher API gravities (between 37° and 42°). Not all light oils are tight oils. The term "shale oil" is also used to refer to tight oil, but is most commonly used to refer to oil derived by pyrolysis from oil shale which is an organic-rich sedimentary rock containing kerogen.

3.2. Growth in U.S./Canadian Rail Transport of Crude

As production of tight oil from U.S. shale formations increases, challenges arise in bringing new production to market. Infrastructure challenges include delivery of materials, equipment, and energy needed to sustain production. However, assuming these challenges are met and production increases, the two main challenges of moving increased production to market are transportation of product and pending regulations.

Almost all oil produced domestically flows to one of the 123 U.S. refineries (Koottungal, 2014), with nearly 45% of the country's refining capacity located along the Gulf Coast. Pipelines and oceangoing tankers still deliver the vast majority of crude to U.S. refineries, accounting for approximately 90% of total receipts (in barrels) in 2013 (Frittelli, 2014a). The first step in moving product to market is getting it from the well site to a distribution point (or refinery). In North Dakota, oil is transported from the well site either by truck (56%) or gathering pipeline (44%) and delivered to rail terminals or pipelines for shipment out of state (Kringstad, 2014). In Texas, oil is delivered to regional refineries by pipeline, truck or barge, and a vast majority of Canadian crude is moved by pipelines (Frittelli, 2014a).

Drilling activities in the Williston Basin have moved from securing leases to the harvesting phase. This involves drilling multiple wells on a single pad, which will lead to a potential increase of gathering pipelines and reduce transportation by truck within the state. Since 2009, there has been a steady increase in the number of miles of pipelines in North Dakota (see Figure 3-1) with 2,578 additional miles added in 2013 (Kringstad, 2014).

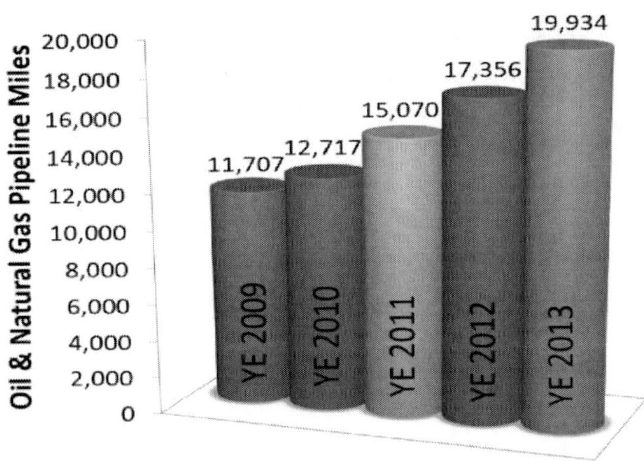

Figure 3-1. North Dakota Oil and Gas Pipeline Miles, Reproduced with Permission from Kringstad (2014).

Oil is transported out of North Dakota by pipeline, rail or truck. It is delivered to refineries or to transfer points where it is loaded onto barges for delivery to refineries. Figure 3-2 plots the estimated split of transportation of oil out of the Williston Basin, and Figure 3-3 shows projected production and export capacities for the Williston Basin.

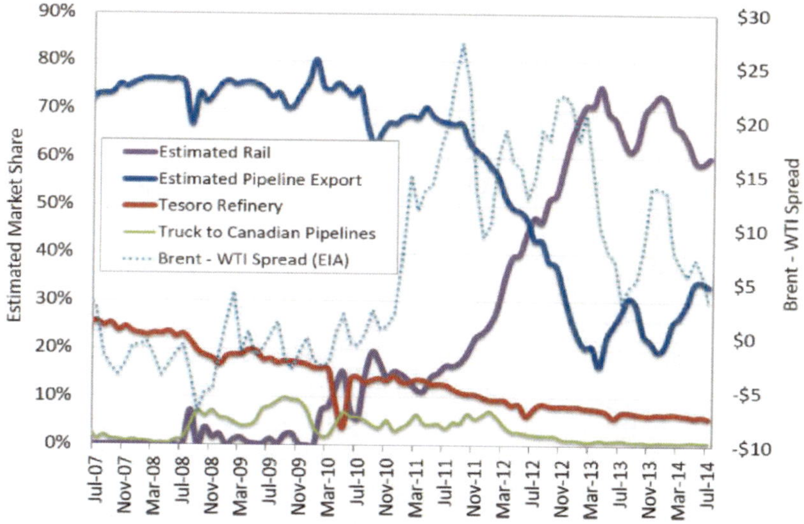

Figure 3-2. Estimated Williston Basin Oil Transportation, Reproduced with Permission from Kringstad (2014).

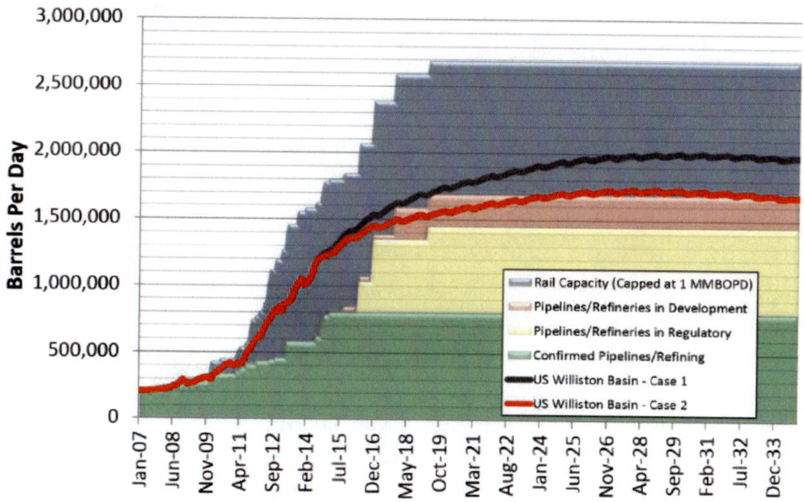

Figure 3-3. Williston Basin Oil Production and Export Capacity, Reproduced with Permission from Kringstad (2014).

In the coming years, it is projected that the capacity to move oil out of the Williston Basin could almost entirely be met by pipelines. However, the final destination (refineries) and the means of getting the oil there will be affected by a number of factors including the following:

- Price of oil
    - Lower crude oil price makes it less attractive to ship by rail to refineries.
    - Higher prices lead to increased production and higher demand for transportation.
    - Discounted prices for Bakken crude make it profitable to ship by rail (Frittelli, 2014a).
- Pipeline bottlenecks – Increased pipeline capacity will make it cheaper to transport to refineries in the Gulf Coast region.
- Reconfiguration of refineries to process light sweet crude – Currently, Gulf Coast refineries are blending the light sweet crude with heavy crude to achieve a medium API gravity crude oil that can be processed at these refineries (DOE, 2014).
- Canadian oil production – It is expected that shipment of Canadian oil by rail will more than triple by 2016 (Cattaneo, 2014).
- Texas oil production – As production from the Permian Basin and Eagle Ford Shale increases there will be less demand for Bakken at Gulf Coast refineries. This may increase shipment by rail to the East and West coast refineries.

Based on this information, it appears industry will continue to increase capacity for transporting not only crude oil from the Williston Basin, but also from Western Canada. If the Keystone XL pipeline is completed, this will significantly increase the capacity for transporting oil from the tar sands region of Canada and relieve congestion in U.S. oil transportation systems (Frittelli, 2014a).

In Texas new pipelines are being built to connect increasing production from the Eagle Ford and Permian Basins (Permian Express Phase I and II, Granite Wash extension, Eaglebine Express, and Permian Longview and Louisiana Extension). These projects have an approximate combined capacity of 520,000 barrels per day. Also, Rangeland Energy is building a 100,000 bbl/d rail loading terminal to transport crude oil from the Permian Basin (EIA, 2014b). If production from the Eagle Ford Shale and Permian Basin exceed Gulf Coast refining capacity, oil may be shipped by barge or tanker to east and west coast refineries cheaper than oil can be delivered from North Dakota by rail. The only obstacle to this scenario is the Merchant Marine Act of 1920 (aka The Jones Act), which requires vessels transporting cargo between two U.S. points be built in the United States, as well as crewed and at least 75% owned by U.S. citizens (46 CFR, 2014). Tankers built in the United States are about four times the price of foreign-built tankers (Frittelli, 2014b). There is also limited capacity at U.S. shipyards to build new tankers.

While industry can build more infrastructure to move product to market, it may take more effort to deal with new regulations on oil transport. Recent events have increased safety and environmental concerns over the transportation of oil by rail. Following the catastrophic fire in Lac Mégantic, Quebec, the U.S. Department of Transportation (DOT) issued a safety alert warning that the type of crude oil being transported from the Bakken region may be more flammable than traditional heavy crude oil (DOT, 2014). In July 2014, DOT issued a Notice of Proposed Rulemaking proposing a range of measures to enhance the safety of shipping crude oil by rail, including proposing enhanced tank car standards, a classification and testing program for mined gases and liquids, and new operational requirements for high-hazard flammable trains that include braking controls and speed restrictions. A companion Advanced Notice of Proposed Rulemaking was also issued seeking further information on expanding comprehensive oil spill response planning requirements in shipments of flammable materials (CFR Title 49). These rules may only temporarily impact shipment of oil by rail because in February 2014, BNSF Railway announced that it would purchase 5,000 tank cars, and as of April 2014, manufacturers had 50,000 crude oil tank cars on order (NTSB, 2014).

The North Dakota Industrial Commission (NDIC) has also approved a policy to reduce flared gas from oil wells (2014a). The policy which was approved on July 1, 2014, will restrict production if a producer fails to meet requirements to capture natural gas at the well site. This policy took effect October 1, 2014, with the goal of reducing flaring to 10% by 2020. This new policy may slow production, which would also free up capacity in existing oil transportation systems.

With the increase of shipment of oil by rail comes concern about safety. As mentioned in the previous section, recent events have served to highlight the dangers of shipping oil by rail. Over the period 1996-2007, railroads consistently spilled less crude oil per ton-mile than trucks or pipelines. However, the data precede the recent dramatic increase in oil transportation by rail. In 2008, 9,500 carloads of oil were delivered by rail compared to 407,761 carloads in 2013 and 229,798 carloads in the first half of 2014 (Association of American Railroads, 2014). The increase in rail traffic is due to insufficient pipeline capacity to carry the increasing production of North Dakota and Canadian crude oil to refineries. Crude imports by rail from Canada have increased more than 20-fold since 2011 (Kringstad, 2014). Because of these factors up to 70% of North Dakota oil is delivered to refineries by train (Esser, 2014). Based on PHMSA data, 1.15 million gallons of crude oil were spilled from rail cars in 2013, but only 800,000 gallons were spilled from 1972 thru 2012. The 2013 data do not include the 1.5 million gallons of spilled in Lac-Mégantic, Quebec. Based on the above data, 99.99% of shipment arrived without incident. A report from the American Association of Railroads' report (AAR, 2014) found that rail transport spills 0.38 gallons of oil per million barrels moved, compared to 0.88 gallons for pipelines.

3.3. Well Site Petroleum Conditioning

Well site conditioning - the processing of crude oil in proximity to the production well to remove impurities prior to transport - has the potential to impact combustion-relevant crude oil properties. Impurities include gases, water and solids that were produced with the crude oil. Gaseous product includes hydrocarbon and inorganic gases that either made up a separate phase exiting the well or that were released from the crude oil during conditioning. The gaseous product has economic value, so it is generally collected in relatively low-pressure gathering lines which convey the gases to gas plants which, then, process the stream and ultimately sell it as natural gas and natural gas liquids or, in the absence of gathering lines, is flared at the well site. Depending on the quality of the produced water, the water may be recycled or treated and disposed of—often by deep-well injection. Solids that were carried with the oil also require treatment and proper disposal.

Minimally, conditioning that is performed at the well site yields a liquid product that can be safely and economically transported to a refinery or to another facility for further processing. Stabilization is a process whereby higher vapor pressure components in the crude oil are removed and marketed separately from the lower volatility oil.

Producing well sites are unmanned facilities which are visited only periodically by persons performing such functions as collecting crude oil or wastes, maintaining equipment and testing or otherwise operating wells. Since well site equipment, for the most part, is unattended, process and equipment reliability is crucial.

Petroleum producers frequently say that no two wells are the same. Consequently, even though there are a limited number of basic types of equipment and configurations involved in crude oil conditioning, the specific configuration and operating conditions adopted for a particular well site depend upon the nature of the producing formation, the location and condition of the well, and ambient conditions at a particular point in time. Time is an important factor in well behavior for many tight oils because the depletion curve, which describes the change in petroleum production rate of a well over time, decreases rapidly over time as exhibited in Figure 3-4.

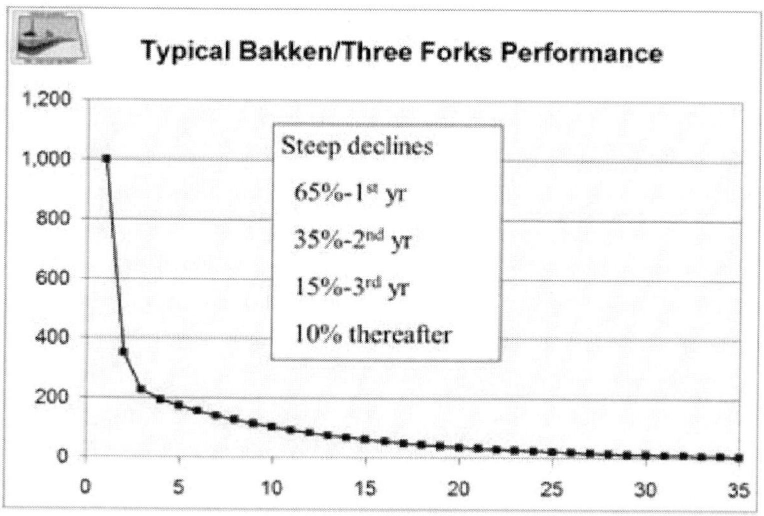

Figure 3-4. Notional Depletion Curve for a Bakken Well, Reproduced from NDIC (2014e).

Ultimately, the key conditioning equipment is a series of gravity-assisted, pressurized multiphase separators. The number, size, and design of these separators depend upon wellhead conditions, rate of production, relative amounts and phases of impurities, and the nature or composition of the petroleum. Different compositions of petroleum possess different affinities for impurities, such as water or sediment. Depending on the composition of the petroleum, for example, the oil and water might easily form two phases with little treatment or the two might form an emulsion that could require chemical additives, heat, quiet flow, centrifugation or other means to separate. After transit through separators, the conditioned petroleum as well as segregated wastewater and solids streams are retained on-site in tanks until collected and transported away from well site by truck or pipeline. Figure 3-5 depicts a notional Bakken well site conditioning configuration (North Dakota Industrial Commission, 2014d). The number of separators, existence of a free-water knockout vessel just upstream of the crude heater and other equipment varies among well sites and over time at a single, specific well site. Figure 3-6 and Figure 3-7 provide pictures of Bakken well sites with equipment notations (Oasis Petroleum, 2014a).

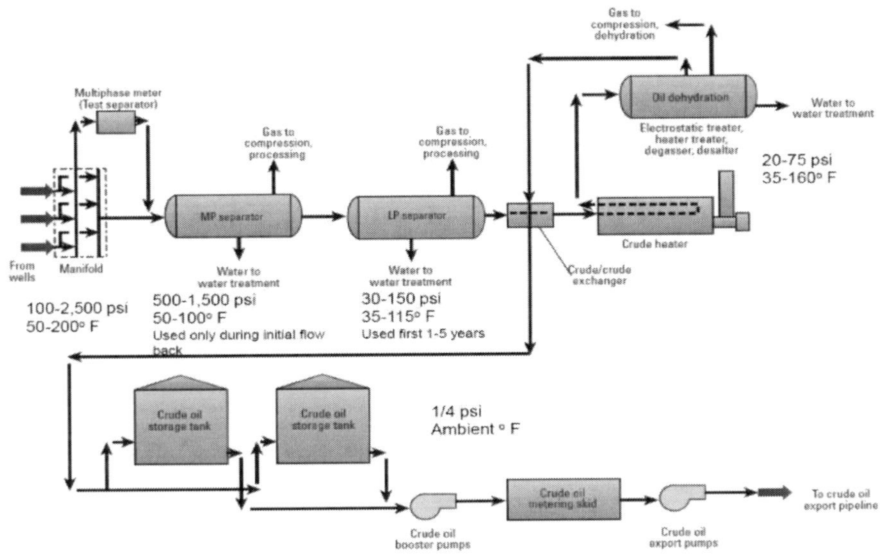

Reproduced from North Dakota Industrial Commission (2014d).

Figure 3-5. Notional Bakken Well Site Conditioning Equipment Configuration.

Reproduced from Oasis Petroleum (2014a).

Figure 3-6. Bakken Well Site Conditioning Equipment Configuration.

Reproduced from Oasis Petroleum (2014a).

Figure 3-7. Bakken Well Site Conditioning Equipment Configuration.

The most common separator reported in the Bakken (Auers, et al., 2014) is the emulsion heater treater (HT). As are most separators, HTs are capable of removing gases, free water, and solids from petroleum; however, their ability to heat petroleum enables them to "break" water-oil emulsions to free and remove water that would otherwise have been tied up and carried with the crude oil.

There are two basic designs of HTs: vertical and horizontal. Horizontal HTs are better able to handle higher flow rates and provide enhanced settling time; however, they must operate at higher pressures than vertical HTs to avoid the need to incorporate a pump to transfer the conditioned oil to storage. The addition of a pump marginally adds complexity and reduces the reliability of the conditioning process. Figure 3-8 and Figure 3-9 are cross-sectional drawings, and Figure 3-10 and Figure 3-11 are pictures of installed HTs.

A study of Bakken petroleum (Auers, et al., 2014) reported a range of HT conditions during sampling, with pressures varying from 12 to 68 psig and temperatures ranging from 35° to 162° F with medians of 36 psig and 94° F, respectively. Whiting Petroleum (2014) has stated that, generally, the HTs that it operates can process two wells' production simultaneously: exceptions are large wells which require that individual HTs be exclusively devoted to each well. It has also stated that, with respect to heating produced oil to reduce its vapor pressure (as opposed to breaking emulsions), vapor pressure and light hydrocarbon content are more dependent upon vessel pressure than temperature. Additionally, it has commented that temperatures above 120° F are "counterproductive" because of the tendency under those conditions to induce salt precipitation which can cause detrimental plugging and fouling of equipment.

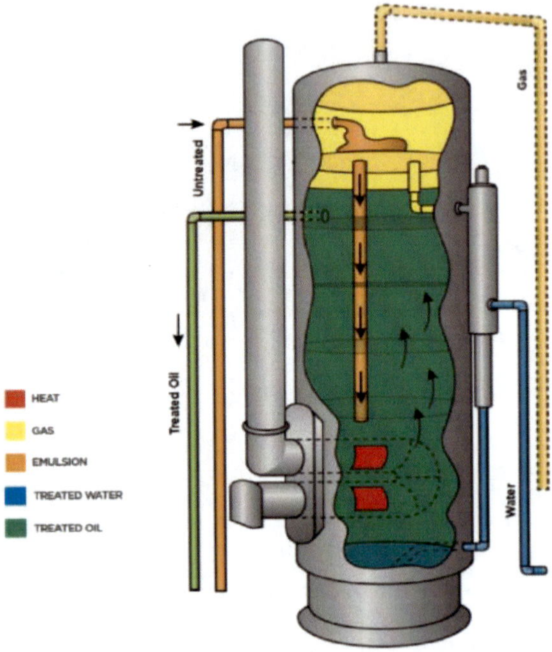

Reproduced with Permission from Diverse Energy Systems (2013).

Figure 3-8. Notional Vertical HT Cross-Sectional Diagram.

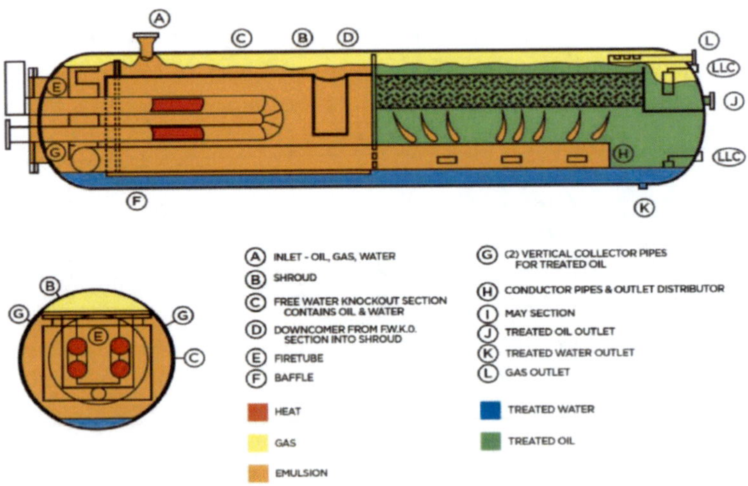

Reproduced with Permission from Diverse Energy Systems (2013).

Figure 3-9. Notional Horizontal HT Cross-Sectional Diagram.

Reproduced from Oasis Petroleum (2014a).

Figure 3-10. Installed Vertical HT.

Reproduced from Oasis Petroleum (2014a).

Figure 3-11. Installed Horizontal and Vertical HTs.

In seeking, "...to improve the transportation safety and crude oil marketability," of tight oils produced in North Dakota, NDIC held a hearing (Case Number 23084, Order Number 25417) on September 23, 2014, to receive testimony regarding regulation of well site crude oil conditioning (NDIC, 2014a). A working draft order was submitted to NDIC on November 13, 2014, which aimed at regulating delivered crude oil at Reid Vapor Pressure (RVP) no greater than 13.7 psig at custody transfer. The approach focused on regulating operation of a well site emulsion HT or other heated-feed gas-liquid separator at conditions listed in Table 3-1 (NDIC, 2014b). Responses received during the draft order's comment period included concern that the required temperatures might negatively impact operator's gas gathering equipment that cannot tolerate fluids heated to more than 120° F (Herr, 2014). Consequently, the final order reduced the required temperature to a minimum of 110° F as indicated in Table 3-2 (North Dakota Industrial Commission, 2014c).

Table 3-1. Draft Order HT and Separator Operating Requirements (NDIC, 2014b)

Equipment and Conditions	Minimum Temperature	Comment
50 psig or less	115° F	
50 to 75 psig	120° F	
More than 75 psig	120° F	Vapor recovery is required on crude oil storage tanks
Other conditions		13.7 psi maximum RVP; quarterly third-party testing
No HT or separator		NDIC approval; 13.7 psi maximum RVP or safe delivery to stabilization plant

Emulsion HTs are potentially more expensive to procure, install and operate than standard, unheated separators because of their greater complexity and heating requirements. To avoid the large size and large amount of heat required to handle wells' total production, unheated separators frequently are installed upstream of HTs to reduce impurities and the volume of material that HTs would otherwise have to process. Such separators remove the bulk of impurities, thereby permitting HTs to process streams that are predominantly emulsions. This is not to say that every well site necessarily has upstream separators—as has been noted, each well is different and conditions change over time; consequently, there are wells whose total production rates or relative amount of impurities are small enough that they can be economically processed by an HT without the need for upstream separations. At this point in the life cycle of the Bakken, such situations appear to be the exception rather than the rule as indicated by a study of Bakken petroleum (Auers, et al., 2014) which reported information describing separators at nine of fifteen Bakken well sites.

Table 3-2. Final Order HT and Separator Operating Requirements (NDIC, 2014c)

Equipment and Conditions	Minimum Temperature	Comment
50 psig or less	110° F	
More than 50 psig	110° F	Vapor recovery is required on or before crude oil storage tanks
Other conditions		VPCRx (ASTM D6377) no greater than 13.7 psi or 1 psi less than the vapor pressure of stabilized crude as defined in ANSI/API RP3000 (whichever is lower) with quarterly testing
No HT or separator		NDIC approval of alternative equipment; VPCRx 13.7 psi maximum or safe delivery to stabilization plant

VPCRx – vapor pressure of crude oil at constant vapor/liquid ratio (V/L), where x=V/L. Test temperature and V/L are not specified in NDIC (2014c).

Similarly to emulsion HTs, separators are pressurized vessels that use gravity to facilitate separation, are designed in vertical or horizontal orientations, and can perform three-phase separation (gas, water, and emulsion or oil). However, in contrast, some separators are designed only to process two phases: gas and emulsion, or water and oil. When multiple separators are connected in series, they are differentiated by their pressures, e.g., "high-pressure," "medium pressure," and "low pressure." The petroleum experiences a sequence of pressure drops as it progresses through a series of separators: each drop in pressure induces lower vapor pressure components in the oil to vaporize and exit the separator in the gas stream; and each separator provides time (i.e., residence time) and conditions that facilitate the movement of these lighter components through and from the oil phase to the gas phase.

Free-water knockout vessels/separators are represented by equipment fabricators to be low-cost solutions to oil-water separation (Accurate, Inc., n.d.). They operate in a lower-pressure regime, typically just upstream of the lowest-pressure separator (i.e., HT), and provide a large oil-water interface to facilitate disengagement and separation. As is true of other separators, these vessels are designed in vertical and horizontal orientations.

Designs and definitions of separators vary among individuals, petroleum production companies and equipment fabricators. For example, Schlumberger, Ltd.'s online "Oilfield Glossary" states that "a free-water knockout is commonly called a three-phase separator because it can separate gas, oil and free water" (2014). This definition is contrary to that of an industry intelligence and consulting firm, the Oil and Gas Mineral Service Company, which is "a vessel used near a well location to separate produced brine from crude oil. These are used generally when there are high volumes of produced water. Also referred to as an oil-water separator" (Oil and Gas Mineral Services, 2011).

Another variation related to free-water knockout vessels is whether or not they are needed at a specific site: as evidenced by Continental Resources (2014) inclusion of free-water knockout in flow diagrams of their Bakken well sites, but an absence of free-water knockout in diagrams and a well site picture of Oasis Petroleum facilities in the Bakken (Oasis Petroleum, 2014a).

When Bakken well site facilities are compared with those of other tight oil production areas, many similarities are apparent. The similarities are also apparent in the equipment list provided by one well site equipment manufacturer, which, for one Eagle Ford project in 2012, included "hi/low separator packages, metering skids, heater treaters, ET H2S scavenger packages, filter separators, TEG dehydration systems, and storage tanks" (Valerus, 2012). However, Eagle Ford production is lighter than Bakken and frequently requires stabilization of the produced oil.

The Marcellus and Utica shale formations have substantial gas which requires a different set of well site equipment. The Eagle Ford central gathering facility, thus, represents a step in a transition from an oil shale play to a gas shale play and, with that transition, presents not only greater complexity both in terms of layout, but also the specific equipment employed to perform the processing.

Well site equipment and layout become increasingly dissimilar to Bakken facilities as reservoir location and type differ from Bakken shale and as technologies employed at the well site differ from hydraulic fracturing. An example of this includes a single train of the Snorre A platform (Norwegian Sea) conditioning system, in which a three-phase separator is placed between two two-phase separators (Kylling, 2009). A second example of such dissimilarity is the application of enhanced oil recovery methods, such as the enhanced oil recovery. In these situations, significant volumes of carbon dioxide are produced and must be recovered for recycling during the later stages of carbon dioxide flooding of a formation (Moritis, 2001).

3.3.1. Stabilization vs. Conditioning

The purpose of conditioning is to remove impurities from crude oil with the intent to eliminate compounds (typically nonhydrocarbon compounds) from the oil that lack fuel value or that unnecessarily elevate the hazard level of the petroleum, e.g., presence of toxic hydrogen sulfide. The principal purpose of stabilization, in contrast to conditioning, is to remove hydrocarbon compounds—which possess fuel value, but which also possess higher vapor pressures (i.e., have lower boiling points)—so as to reduce the volatility of the crude oil. It should be noted, however, that stabilization also can remove remaining traces of higher vapor pressure impurities remaining in the oil. Even though the two processes have similar effects, i.e., removing components from crude oil, they are quite different in the effort and equipment required to perform the separation because, in a sense, water and inorganic gas impurities have inherent propensities to form separate phases from hydrocarbons in crude oil which aids separation, whereas lighter (lower molecular weight) hydrocarbons which are the major contributor to crude oil volatility, have inherent affinity for the oil which impedes separation. The impact of this difference is that carefully designed fluid motion and step drops in pressure using gravity and available wellhead pressure, as well as possible chemical and heat addition, can be produced and introduced by equipment of relatively simple and reliable design with limited energy requirements to accomplish satisfactory conditioning. Stabilization, on the other hand, typically requires more complex processing and equipment, as well as more rigorous, energy-consuming vaporization and condensation of the crude oil. Additionally, while stabilization produces a lower vapor pressure crude oil, it also produces a vapor product that contains higher boiling constituents than those produced from its associated upstream conditioning system. Because of their tendency to condense in pipelines, their greater energy density in the vapor state and other reasons, natural gas pipeline operators have restricted how much of these constituents can be introduced into a natural gas pipeline. Consequently, the lighter-components stream produced by stabilization requires 1) a process to remove components that are not permitted to be transported in natural gas pipelines from those that are permitted and 2) a transportation system which is separate from both natural gas and stabilized crude oil.

Fractionation by means of a distillation column is the conventional means of performing crude oil stabilization. In this process, conditioned crude oil is fed into a vertical column containing carefully fabricated structures that enhance vapor and liquid contacting to promote mixing of vapor and liquid such that more volatile components in the oil transfer to the vapor phase while the less volatile condense in the liquid phase. Heating of the liquid in a reboiler at the bottom of the column promotes movement of more volatile components in the oil to vaporize and move upward in the column, exiting at the top as a vapor. The complexity of the design and equipment makes the process more expensive and less reliable and requires more attention than simpler conditioning processes. Figure 3-12 depicts this process: associated equipment—such as a furnace to heat the reboiler heating medium, a source of cooling medium for the crude oil cooler and a cooler for the vapor product, as well as ancillary pumps and other equipment—is not included in the figure, nor is the gas-processing system that removes constituents that cannot be introduced into natural gas pipelines.

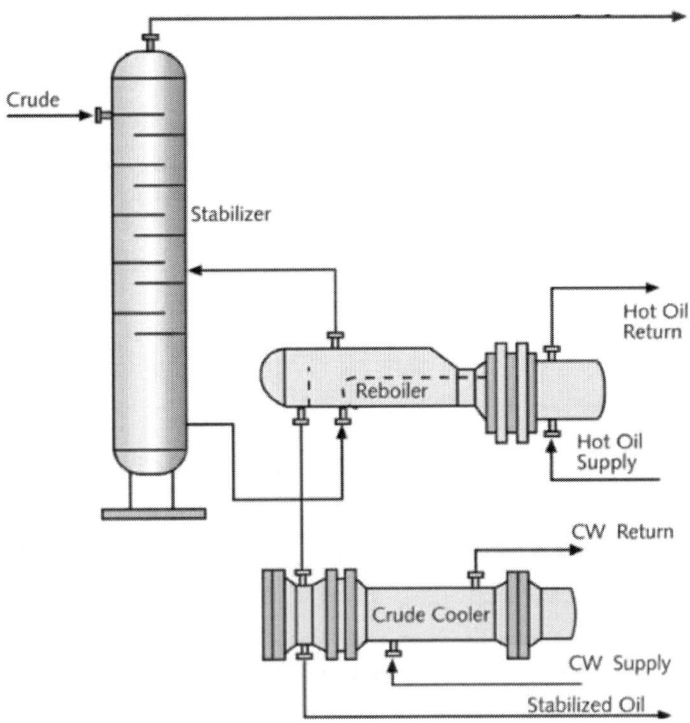

Reproduced with Permission from Gardner Denver Nash (2008).

Figure 3-12. Notional Equipment Layout of a Crude Oil Stabilizer.

Another approach to petroleum stabilization involves multistage flash. Similarly to conditioning processes, multistage flash processes can be installed at well sites; however, they differ from conventional conditioning processes by possessing more stages that make more efficient use of pressure and heat management. Consequently, they require more engineering effort and more fabrication and likely have more restricted operating regimes than conventional conditioning equipment. An example of a horizontal staged-flash stabilizer installed at Eagle Ford is shown by Valerus (2014a). Another example, a 2,500 barrel per day dual-train unit installed in 2012, is shown for a Utica well site in Valerus (2014b). This approach appears to be relatively new; this study was unable to discover applications of the technology prior to 2012 or reports of the effectiveness of the equipment as production rates declined over time.

4. Basic Description of Crude Oils

Crude oils comprise a complex mixture of primarily liquid hydrocarbons with dissolved gases and trace amounts of suspended water and inorganic sediments. API (2011) indicates that from an elemental standpoint, an average crude oil contains the following:

- 84% carbon
- 14% hydrogen
- 1-3% sulfur
- 1% nitrogen
- 1% oxygen
- 0.1% minerals and salts

The organic compounds in crudes are primarily straight and branched-chain alkanes, cycloalkanes, cycloparaffins, and aromatics. Composition of crude oils varies widely across producing regions, and even within given geologic formations. Composition and properties can also change within a given stream as the oil moves from the reservoir to the refinery as a result of sequential conditioning, stabilization and commingling steps. The discussion below is a targeted discussion addressing selected oil properties useful to general description along with some properties pertinent to safety in handling and transport.

4.1. API Gravity and Sulfur

The common designation of “heavy” or “light” crude is based on its density. API gravity is a common unit of measure, calculated as shown in Equation 4-1.

$$°API = \left(\frac{141.5}{RD}\right) - 131.5 \tag{4-1}$$

where *RD* = relative density (as compared to pure water at 60° F). Lighter oils exhibit higher API gravities, while heavier oils exhibit lower. An example conversion table is shown below (Table 4-1) for API gravity, absolute density, and relative density.

Note as a reference, pure water exhibits an API gravity of 10 degrees. While light and heavy are relative terms, there is some level of acceptance across the industry that “light” crude oils exhibit ≥ 33°API, while “heavy” crude oils exhibit ≤ 28°API (API, 2011).

The designation of “sweet” or “sour” oil relates to the total mass of sulfur. Definition of sweet versus sour varies by region and context. API (2011) cites a general rule that crude oils with less than 1% total mass sulfur are sweet, while those with greater than 1% are sour. The U.S. Strategic Petroleum Reserve (SPR) defines sweet oils as having total sulfur mass% < 0.5, and sour having total sulfur mass% < 1.99. Sulfur is an impurity that must be separated from crude during the refining process. Sulfur is also problematic because it is associated with components of oil that are corrosive, volatile, poisonous, and flammable.

A 2014 study commissioned by the North Dakota Petroleum Council (NDPC) (Auers, et al., 2014) analyzed Bakken oil samples and compared their results with important domestic and international crudes for API gravity and sulfur content, summarized below in Table 4-2.

The Bakken and Eagle Ford oils, which are produced from “tight” formations, comprise the upper end of the API range shown, ranging from 40-58.8°API. Sulfur content for the tight oils is at the low end, ranging from 0.04-0.1 wt%.

Table 4-1. Example conversions among API gravity, absolute density (RHO), and relative density (RD)

API	RHO	RD
[deg]	[kg/m^3]	[-]
50	779	0.780
45	801	0.802
40	824	0.825
35	849	0.850
30	875	0.876
25	903	0.904
20	933	0.934
15	965	0.966
10	999	1.000

Table 4-2. Comparison of API Gravity and Sulfur Content of Selected Domestic and International Crude Oils, (Auers, et al., 2014)

Crude Description	**API Gravity**	**Sulfur (wt%)**
Domestic Light Sweet Crudes		
Bakken	40-43	0.1
West Texas Intermediate	37-42	0.42
Louisiana Light Sweet	36-40	0.39
Eagle Ford	47.7	0.1
Eagle Ford Light	58.8	0.04
International Crudes		
Light Sweet		
Brent	37-39	0.4
Medium		
Arabian Light	33	1.98
Arabian Heavy	27.7	2.99
Heavy		
Western Canadian Select Heavy Sour	21.3	3.46
Dalia	23.1	0.51

4.2. Crude Oil Phase Behavior

4.2.1. Relevance of Crude Oil Phase Behavior to Handling and Fire Safety

A primary objective for this study is to gain an understanding of the fire and explosion hazards associated with accidental release of crude oil in the transport environment. It is the vapor emissions from a "flammable liquid" that actually burn, so attention is directed here toward the potential for liquid crude oil to volatilize to form a vapor phase in the vicinity of the liquid. Stabilized crude is typically tested, transported, and sold in the liquid phase and associated vapor losses during handling and transport, if any, are not well-characterized. Vapor losses, if any, may not cause measurable financial impact from a sales perspective but could lead to elevated risk from a hazards perspective. It is therefore prudent to examine the phase behavior of crude, specifically the potential for formation of vapor phase emissions, in order to understand the conditions that contribute to fire and explosion hazards around spills.

4.2.2. Definition of Phase Behavior

Phase behavior describes what phases (solid, liquid, or gas) are present under certain thermodynamic conditions of pressure, specific volume, and temperature. Crude oil can also exhibit complex phase behavior with multiple solid phases or supercritical fluids (Mansoori, 2009). Crude oils relevant to this study at ambient temperature and pressure exist largely in the liquid phase, and may produce a co-existing gas phase. Aforementioned solids such as asphaltenes and waxes may also be present, but they tend to be less prevalent in lighter oils and are currently assumed irrelevant to fire risks, though they can create problems with flow assurance.

In the event that a crude oil is involved in a fire, the parameter space for phase behavior must be expanded beyond ambient pressure and temperature conditions common to stock tanks, trucks, and railcars. Oil in a pool fire will experience gradual heating and may volatilize a spectrum of components, favoring those with boiling point temperatures at or lower than the pool temperature. Alternatively, oil remaining in pressurized vessels such as closed railcars impinged by pool fire will experience increases in both temperature and pressure, up to the relief valve and/or burst pressure of the railcar in the specific scenario. The evolution of vapors and time-dependent compositions in the gas and liquid phases in these two cases may follow very different pathways from the same starting "parent" oil.

4.2.3. Pure Substance Phase Diagram

It is instructive to start with a discussion of pure substance phase behavior because pure substances compose the building blocks for mixtures and also provide useful background for the hazards classifications introduced in Section 5.2 An example equilibrium phase diagram of a representative pure alkane is shown in Figure 4-1. The phase diagram relates pressure, temperature and specific volume with a 3-D surface. Such diagrams may be developed from experimental data and/or equation of state (EOS) models. The main value in the phase diagram is in illustrating that there is a unique and predictable thermodynamic state (i.e., specific volume) for a given combination of pressure and temperature.

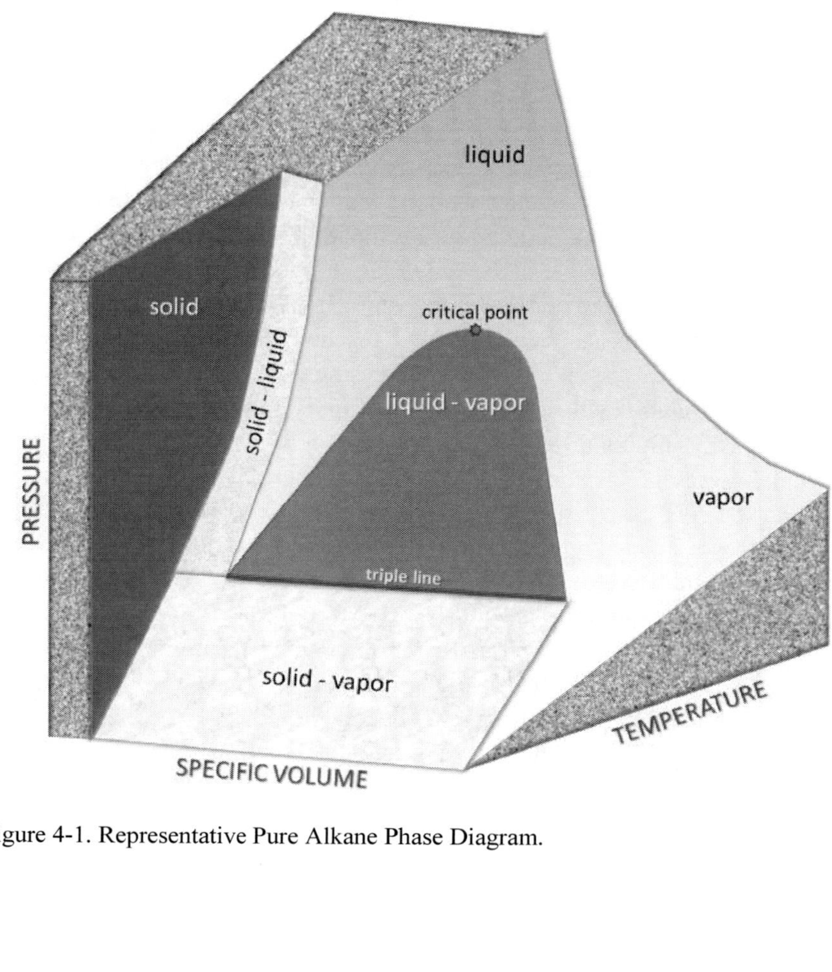

Figure 4-1. Representative Pure Alkane Phase Diagram.

4.2.4. Vapor-Liquid Equilibrium (VLE)

This section focuses on a specific region of the phase diagram, particularly the area where liquid and vapor phases are in equilibrium (annotated as liquid-vapor region in Figure 4-1), as this is the area most instructive toward understanding what conditions promote co-existence of gas and liquid phases. Phase equilibrium is achieved when the chemical potential of each component is equalized across the phase boundary, and there is no net transfer of vapor to liquid or liquid to vapor. It is important to acknowledge that real-world handling, transport, and accident scenarios involving vapor-liquid systems are dynamic, and rarely at true thermodynamic equilibrium. All the aforementioned systems will spontaneously tend toward an equilibrium state, however, so understanding the drivers to equilibrium is critical to understanding anticipated system behavior.

Turning the 3-D surface in Figure 4-1 to look down the volume axis gives a pressure-temperature projection plot shown in Figure 4-2. The curve in Figure 4-2 connecting the triple point with the critical point is where vapor and liquid co-exist. Viewed from this perspective, the VLE condition appears an unlikely situation for so much attention, but many petroleum hydrocarbon systems sit right on this curve during handling, transport, and storage. Generally speaking, this type of system will exhibit a predictable pressure that is a direct function of temperature on the vapor-liquid equilibrium curve over a wide range of fill levels. If the container reaches 100% liquid-full or 100% gas-full, then the system is considered single-phase and the pressure-temperature behavior will depart from the vapor-liquid curve in Figure 4-2. This is all very important information relevant to evaluating the ability of a containment vessel to perform properly under anticipated operating conditions. It also has bearing on whether the fluid system will generate vapors by boiling during a loss of containment when it is subjected to ambient pressure and temperature conditions.

Vapor-liquid equilibria are observed for pure substances as well as mixtures, and for closed as well as flowing systems. A familiar example of a hydrocarbon fuel stored at VLE is the liquefied petroleum gas (LPG) cylinder that contains a mixture of propane + butane liquid and gas. Containment pressure is reasonably predictable and within a narrow band for a given storage temperature, and will not vary much with fill level until the cylinder is 100% liquid-full or 100% gas-full.

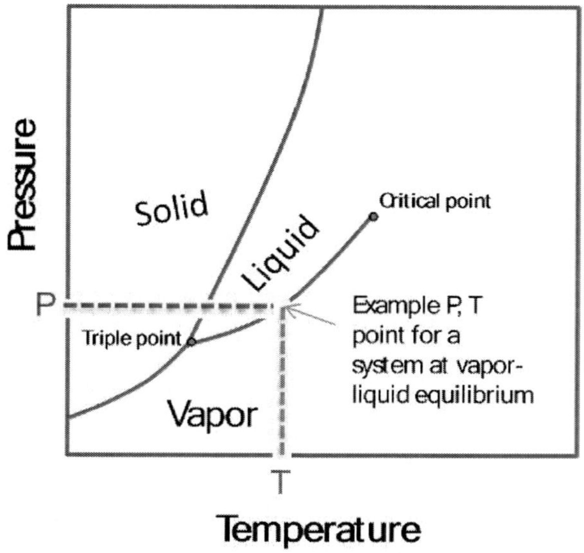

Figure 4-2. Pressure-Temperature Phase Diagram for a Representative Alkane.

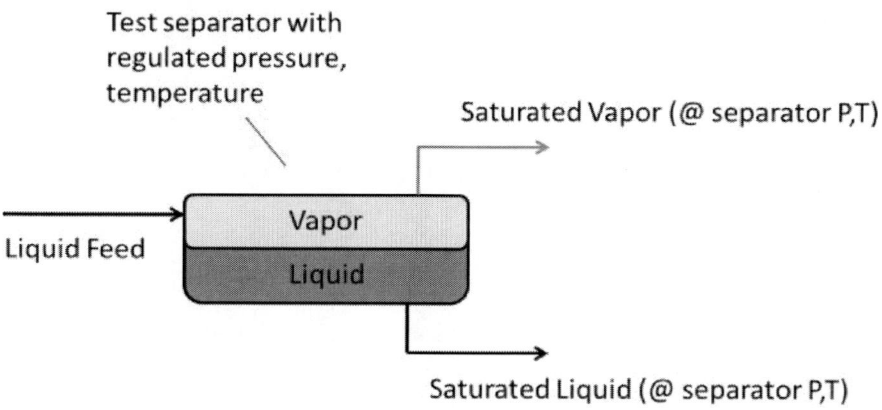

Figure 4-3. Simple Schematic Drawing of a Test Separator with Liquid Feed Entering a Flash Process by which the Feed is Separated into Saturated Gas and Liquid Streams.

A common unit operation used to evaluate VLE properties in petroleum fluids process environments is a test separator, shown in a simple schematic in Figure 4-3. A liquid feed stream is passed through a test separator at a regulated pressure and temperature that promotes phase separation. The produced gas and liquid are collected downstream of the separator. Properly instrumented for flow rates, temperature, pressure, and material stream compositions, this device can obtain useful VLE data for the feed stream. While simple in concept, the design and operation of test separator for acquisition of accurate VLE data requires meticulous attention to detail.

4.2.5. Pure Substance vs. Mixture Behavior

Since crude oil is a hydrocarbon mixture, it will exhibit more complex phase behavior than any one of its single underlying components. Crude oils exhibit a range of boiling temperatures at any given pressure, as opposed to a single boiling temperature exhibited by a pure substance. An illustration of the range of boiling temperatures exhibited for a Bakken sample received by pipeline at the U.S. SPR in December 2012 is shown in Figure 4-4. Note there is a stepwise increase in mass% volatilized with each successive increase in cut temperature.

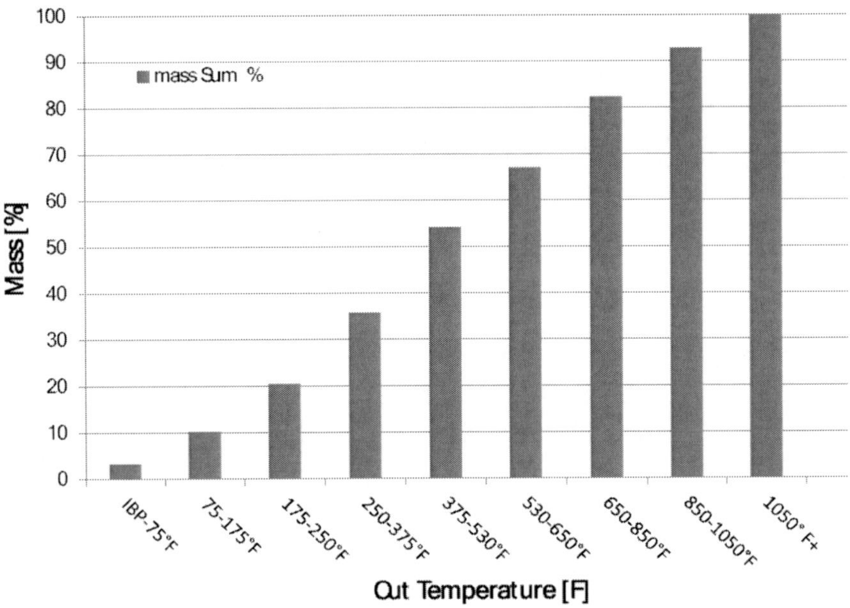

Figure 4-4. Mass % vs. Cut Temperature for Distillation Analysis of a 41.4 API Gravity Bakken Sample Received by Pipeline at the SPR in December, 2012.

Mixtures are also distinct from pure substances in their pressure-volume relationship at vapor-liquid equilibrium. An illustration of the differences is shown in Figure 4-5, where a constant composition expansion (CCE) test is depicted. The test fluid is contained in a piston-cylinder device with just enough pressure to maintain 100% liquid phase, indicated on the left axis of the plot where vapor-liquid (V/L) = 0. As the piston is raised to incrementally increase the containment volume, indicated by moving to the right on the figure, vapor formation is induced and an equilibrium vapor pressure is observed. Total system mass, composition, and temperature are held constant during the expansion. A series of such expansions is conducted at selected V/L values. The critical difference between pure substance and mixture in this test sequence is that the pure substance will maintain a constant vapor pressure at all V/L, while a mixture will exhibit decreasing vapor pressure with increasing V/L. This is shown qualitatively on the figure by contrasting the pure substance curve with the mixture curve.

The bubblepoint pressure (BPP) of the mixture is a special case of vapor pressure where V/L = 0. Any incremental decrease in pressure or increase in expansion ratio from that point will create a gas space in the sample container. Bubblepoint pressure is difficult to measure directly, but may be inferred from P-V measurements in the two-phase region near V/L = 0 coupled with a curve fit to determine the intersection point with V/L = 0 (Feller and McDonald, 1950; Lord and Rudeen, 2010).

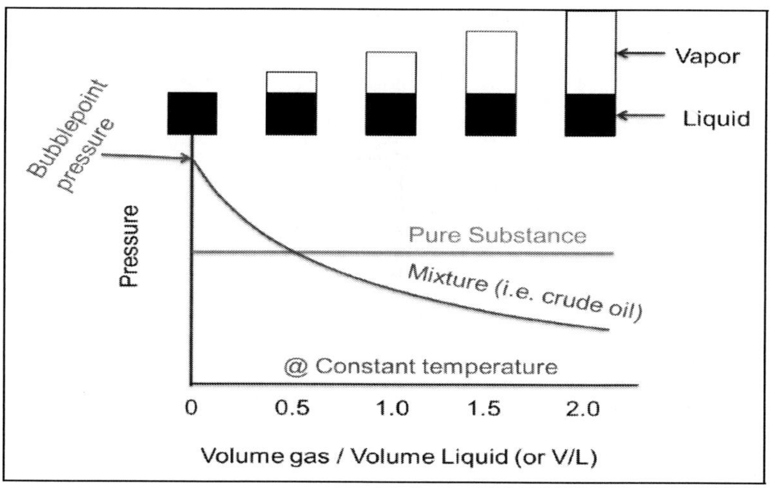

The bubblepoint pressure of a mixture is a special case of vapor pressure defined where V/L = 0.

Figure 4-5. Illustration of Pure Substance and Mixture Behavior for Vapor Pressure vs. V/L at VLE.

4.2.6. Flash Gas Properties

The vapor space created above a liquid in a closed system at equilibrium will contain gases in direct proportion to their relative volatilities and mole fractions in the liquid phase. The gas composition and mole fractions can be measured experimentally or estimated with an EOS model. How much gas and what components are present will have important bearing on the potential hazards realized a release scenario.

SPR has obtained many test separator samples for their crude oils in storage using the basic test separator technique illustrated in Figure 4-3. A representative SPR sweet crude oil, WH108, with API gravity 37.2 was tested in a flash separator in May 2011, giving the following results for flash gas composition at the bubblepoint pressure of 18.1 psia shown in Figure 4-6. The separator was run with a shut-in vapor space and careful control of the liquid-vapor level to achieve an operating point near the bubblepoint pressure of the feed stream at T = 100° F. Note the presence of inorganic gases N_2 and CO_2 together comprising about 27% the mole fraction, with the remaining gas consisting of light hydrocarbons. The peak component in this bubblepoint flash gas is propane at 26%.

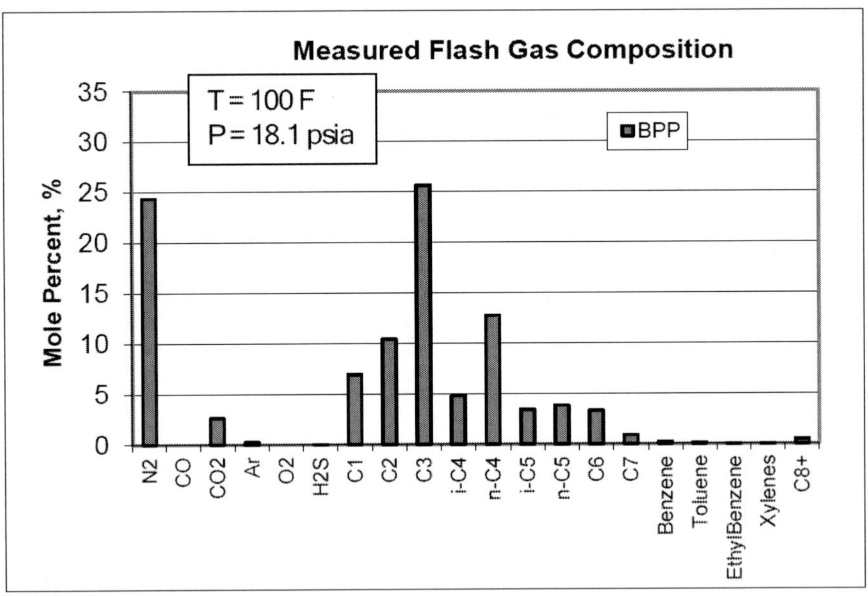

Figure 4-6. Composition of Flash Gas for SPR Cavern WH108 on 5/20/2011 Measured at the Bubblepoint Pressure = 18.1 psia.

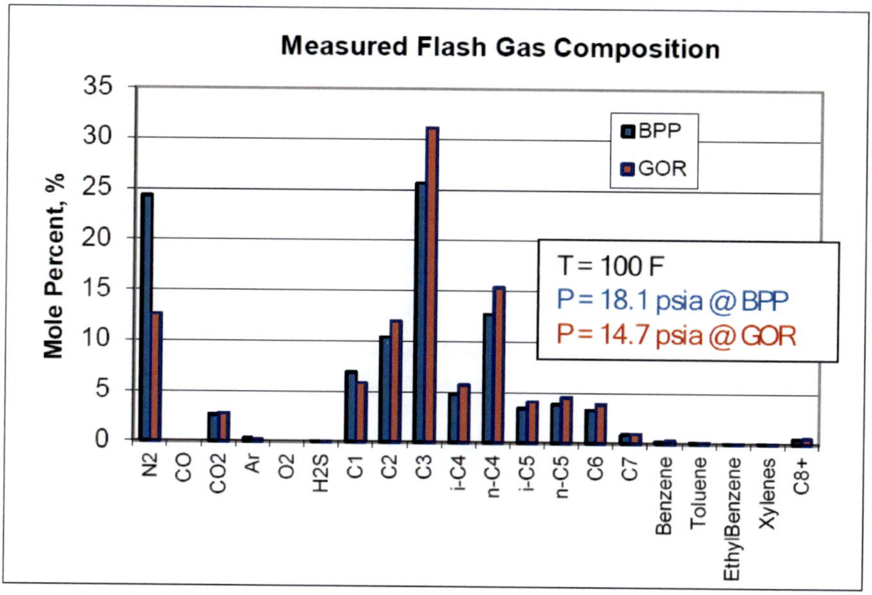

Figure 4-7. Composition of Flash Gas for SPR Cavern WH108 on 5/20/2011 Measured at the Bubblepoint Pressure = 18.1 psia and at 1 atm.

Following the completion of the bubblepoint pressure test, the test separator was de-pressured to atmospheric pressure and allowed to reach a new equilibrium state to measure the gas-oil ratio at the same T = 100° F. The new GOR test data are overlaid with the BPP test data in Figure 4-7, showing how the compositional profile responds to the reduction in separator pressure. Measured GOR for this test was 1.4 scf/bbl. Typical for this type of test on SPR oil, the atmospheric pressure GOR test sees a reduction in N_2 and C_1 mole%, with an offsetting increase in C_2 and heavier hydrocarbons. Note the mole% must sum to 100, so offsetting across the spectrum of measured components is required. As a general statement, vapor space composition above crude oils in near-ambient storage conditions will change to favor incrementally heavier components with decrease in confining pressure, increase in V/L, and increase in temperature.

4.2.7. Constant Composition Expansion Data for SPR Crude

Selected SPR crudes have also been tested using a CCE device via the principle illustrated in Figure 4-5. Expansion points were set to 0.05, 0.2, 0.5, 1.0, and 1.5 at three different temperatures. The instrument could only run three expansion points at one temperature for each sample, so the data shown in Figure 4-8 were generated from aggregate data collected on a steady feed stream of SPR crude over a two-day period (Lord, et al., 2007). The BPP values plotted on the left axis where expansion ratio = 0 were calculated using a curve fit to the V/L = 0.05, 0.2, 0.5 measured expansion points. Note that increasing the system temperature shifts the data upward. At SPR, the general observation is about 1.5 psi gained in vapor pressure for every 10° F increase in temperature in the normal operating range of ~80-120° F.

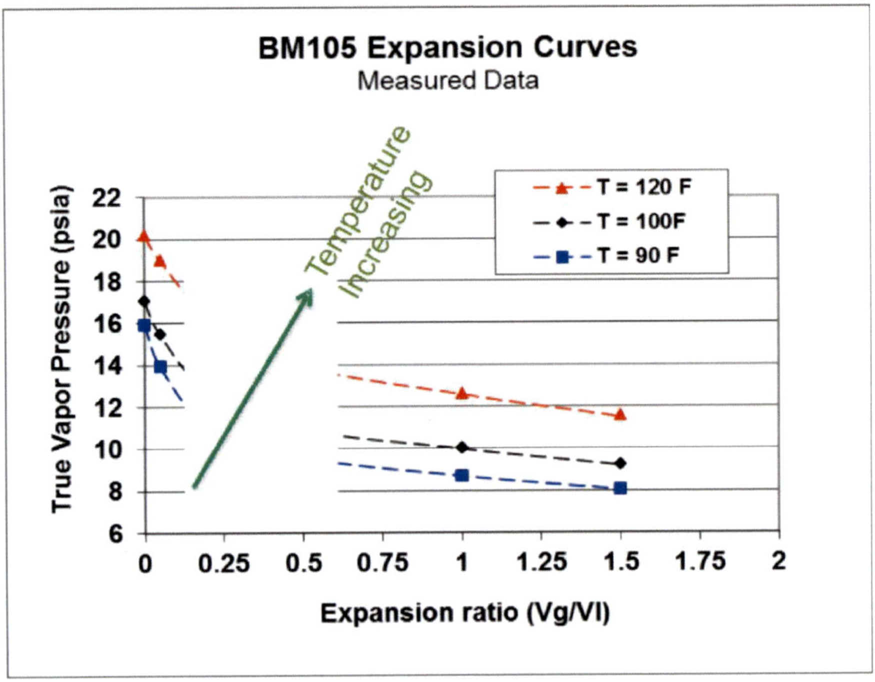

Figure 4-8. CCE Data Measured on SPR oil BM105 in October, 2007.

5. Current Dot Classification Criteria

A brief review of current hazards classification methods is provided in this section to offer context for the ongoing research in crude oil properties as they relate to safety in handling and transport.

5.1. Department of Transportation Authority

The U.S. DOT's PHMSA regulates the transport of hazardous materials by all modes within the U.S. The Hazardous Materials Regulations (HMR) require the shipper to properly classify hazardous material prior to offering it for transportation. In three basic steps, the shipper of a hazardous material must (PHMSA, 2014):

1) Properly identify all the hazards of the material
2) Determine which of the nine hazard classes characterizes the hazards associated with the material
3) Assign each material to a packing group, if applicable

5.2. Hazard Classes

HMR has nine hazard classes, and most crude oils fall into Class 3, flammable liquids. A flammable liquid is defined in this context as "...a liquid having a flash point not more than 140° F, or any material in a liquid phase with a flash point at or above 100° F that is intentionally heated and offered for transportation or transported above its flash point in bulk packaging" (PHMSA, 2014).

Other possible hazard classes for petroleum hydrocarbons include flammable gas (Class 2) or non-hazardous (ANSI/API, 2014). These are not discussed further here.

5.3. Packing Groups

Once assigned to a hazard class, materials may then be assigned to packing groups (PG), from great danger (PG I) to minor danger (PG III). The packing group classification thresholds for flammable liquids (Class 3) are listed below in Table 5-1. The combined results of a flash point and initial boiling point (IBP) test are required for PG assignment. There is an exception if IBP ≤ 95°F in which case PG I is assigned.

Table 5-1. Hazardous Materials Packing Groups Table (PHMSA, 2014)

Packing Group	Flash Point	Initial Boiling Point
I (Great Danger)		≤95°F
II (Medium Danger)	<73° F	>95°F
III (Minor Danger)	≥23° C, ≤60° C (140°F)	>95°F

5.3.1. Flash Point

Flash point is defined by the American Society for Testing and Materials (ASTM) as "the lowest temperature corrected to a barometric pressure of 101.3 kPa, at which application of an ignition source causes the vapors of a specimen of the sample to ignite under specified conditions of the test" (ASTM, 2013b).

Several test methods are allowable, including ASTM D56-05 (ASTM, 2010) and ASTM D93 (ASTM, 2013b). Authors have criticized both aforementioned methods as problematic because light ends may be lost during the sample collection and transfer processes (GPAC, 2014).

5.3.2. Initial Boiling Point

Initial boiling point is typically tested by ASTM D86 (ASTM, 2012), and returns a temperature relating to the first appearance of condensed liquid from a distillation process.

Many authors criticize D86 applied to crude oils as giving widely varying (Auers, et al., 2014), or inaccurate results (GPAC, 2014) or not the most appropriate test method (ANSI/API, 2014) for IBP determination for packing group determination. At issue is that ASTM D86 IBP was not developed for wide boiling range materials like crude oils, and impact of the light ends (C1-C4) will be missed by this method. As a result, the true initial boiling point of the liquid will not be measured, leading to higher IBP's reported than are actually characteristic of the substance. ANSI/API (2014) offers alternative best practices for IBP determination by analyzing for the temperature at which the first 0.5 wt% is eluted using ASTM D7900 (ASTM, 2013a) for boiling range distribution with specific instructions for sample handling, to include closed-container methods for sample acquisition.

6. Comparisons of Property Characterizations in Literature

6.1. Sampling Techniques

The procedure that is required to determine any physical property or composition of a volatile liquid from a process system requires the following steps:

1) Determination of proper location in process to acquire a representative sample for analysis.
2) Determination of proper sampling technique to maintain integrity of the sample, while it is being transferred from the process to the sample container.
3) Determination of proper sample container to maintain sample integrity during sampling, handling, transportation, and delivery of sample to laboratory for analysis.
4) Review and understanding of how sample is to be transferred from the sample container to the analyzer. And determination if the transfer method will compromise the integrity of the sample prior to analysis.
5) Determination of the proper analysis technique required to determine the property or composition of interest.

Often the majority of the emphasis in this effort is focused upon the analysis technique. Usually little review is performed upon the sampling location, sampling technique to be utilized in the field or sampling handling, and sample delivery to the analyzer in the laboratory. Often the usual sampling method is used without regard for the variability in the liquids being sampled and the potential for loss of sample integrity.

6.1.1. Definitions of Open and Closed sampling

Open Sampling: Refers to liquid samples being exposed to atmosphere during sample acquisition into a sample container. *(Typical example is the filling of an open top bottle with sample from a storage tank or process line.)*

Closed Sampling: Refers to the liquid sample being fully contained via tubing, hard piping, etc., directly to a sample container capable of maintaining the sample in a closed, pressure containing, sample cylinder. *(Typical example is a floating piston cylinder (FPC) connected directly to process equipment, with sampling manifold).*

6.1.2. ASTM Manual Sampling Standards Typically Used for Crude Oils

The ASTM has three measurement standards which are routinely utilized in industry to acquire crude oil samples:

- ASTM D4057: Standard Practice for Manual Sampling of Petroleum and Petroleum Products
- ASTM D5842: Standard Practice for Sampling and Handling of Fuels for Volatility Measurement
- ASTM D3700: Standard Practice for Obtaining LPG Samples Using a Floating Piston Cylinder

6.1.2.1. ASTM D4057: Standard Practice for Manual Sampling of Petroleum and Petroleum Products

ASTM D4057 is the sampling standard that is referenced in ASTM D323 Standard Test for Vapor Pressure of Petroleum Products (Reid Method), and is commonly utilized in industry today for acquiring spot or grab samples in the field. The D4057 Standard stipulates the use of various "open type" sampling containers including glass or plastic bottles and cans (ASTM, 2014e). Since D4057 uses open type sampling, the ASTM D4057 Scope states

> 1.1 This practice covers procedures for manually obtaining representative samples of petroleum products of a liquid, semi-liquid, or solid state whose vapor pressure at ambient conditions is below 101 kPa (14.7 psia). If sampling is for the precise determination of volatility, use Practice D5842 (API *MPMS* Chapter 8.4) in conjunction with this practice (ASTM, 2014a).

ASTM D4057 therefore is applicable for the sampling of "dead crude oils" or those which have a vapor pressure of less than atmospheric pressure at ambient temperatures. Even though ASTM D4057 is referenced in ASTM D323 Reid Vapor Pressure Standard, paragraph 10.3 of D4057 states: "10.3 *Vapor Pressure*—When sampling petroleum and petroleum products that are to be tested for vapor pressure, refer to Practice D5842 (API *MPMS* Chapter 8.4)." (ASTM, 2014e)

The D323 Reid Vapor Pressure Standard citation of ASTM D4057 sampling can mislead those who are unaware that D4057 recommends D5842 for vapor pressure measurements.

6.1.2.2. ASTM D5842: Standard Practice for Sampling and Handling of Fuels for Volatility Measurement

ASTM D5842 is another "open sampling" method which is applicable for volatile fuels with a dry vapor pressure equivalent (DVPE) of 2-16 psia (13-105 kPa) (ASTM, 2014c). The sampling methods in this standard are similar to D4057 methods with the exception that for "Reid equivalent vapor pressure" samples, the sample container must be chilled to the warmer of either the liquid temperature in the tank or 32° F (0° C). The D5842 Standard also stipulates that prior to opening the sample container at the laboratory, the sample must be chilled to 32-34° F (0-1° C) and that any vapor pressure testing shall performed on initial material removed from the sample. The remainder of the sample in this container may not be used for vapor pressure testing, but could be utilized for other testing. It should be noted that this standard was developed for light and reformulated fuels which, typically, do not contain significant amounts of light hydrocarbons (methane-butane). And again, is not suitable for sampling of "live" oils.

6.1.2.3. ASTM D3700: Standard Practice for Obtaining LPG Samples Using a Floating Piston Cylinder

ASTM D3700 is a "closed sampling" method that was developed for sampling highly volatile LPG, the vapor pressure of some of these LPG streams are greater than 500 psi (ASTM, 2014b). The D3700 Standard uses a floating internal piston in a high pressure sample cylinder to effectively provide two isolated compartments within the cylinder. Figure 6-1 below provides a conceptual diagram of a typical floating piston cylinder. The top side (pre-charge side) is pressurized to minimum of 10 psig above sample point pressure, with an inert gas. The sample (product) inlet valve, on bottom side of cylinder, is connected to the sample point with tubing or high pressure hose to allow sampling of high pressure LPG.

Once the sample line is connected and sample valve is opened the pre-charge gas is slowly released, lowering the pre-charge chamber slightly pressure below the sample line pressure and sample begins to fill the lower sample chamber, moving the piston upward. The design of an FPC requires the sample source to provide enough pressure to move the internal piston. The piston is designed with o-ring or other type of seal material to isolate the two chambers within the cylinder. This piston seal provides friction to inhibit free movement of the piston. In LPG sampling the sample process pressure is high enough to easily move the piston. For crude oil sampling, the typical sample points are atmospheric tanks which are at very low head pressures. This makes utilizing the typical FPC impractical without a sampling pump or other means to raise the sample pressure. This issue was cited in the NDPC study (Auers, et al., 2014) as a limiting factor to taking crude oil field samples with FPC's.

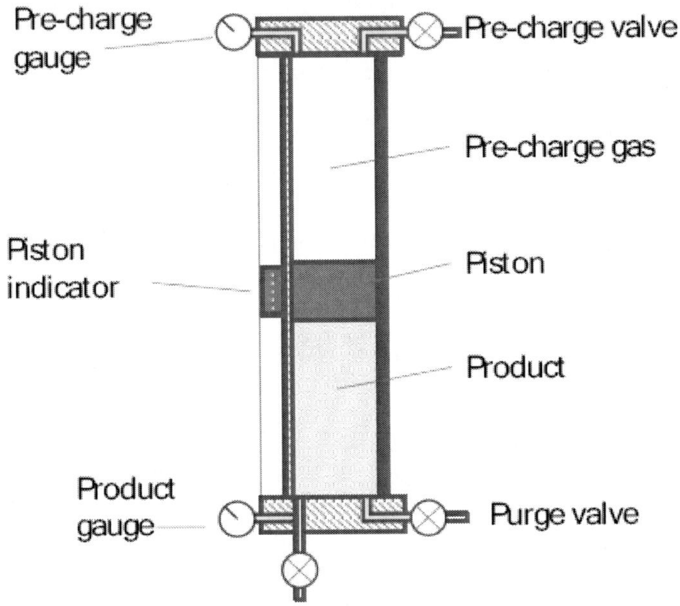

Figure 6-1. Conceptual Diagram of a Typical Floating Piston Cylinder.

6.1.3. Studies Recommending Closed Sampling for Vapor Pressure Determination

There have been several studies to determine the impact of utilizing open vs. closed (pressurized) sampling techniques upon vapor pressure measurements of crude oils. Studies by Pichler and Lutz (2014) and Konecnik (2014) (Figure 6-2), both conclude that in side by side comparison of open vs. closed sampling the VPCR4 (37.8° C) crude oil vapor pressures from the closed sampling procedure (FPC) were consistently higher than the open sampling results. This has long been acknowledged to be correct for "live crude oils." However, these two studies included crude oils with VPCR4 (37.8° C) ranging from 40-120 kPa (6-17 psi), with the majority of the samples being "dead crude oils" with VPCR4 (37.8° C) ranging from 50-70 kPa (7.5-10 psi). The NDPC (Auers, et al., 2014) attempted to study the impacts of open vs. closed sampling upon VPCR4 (37.8° C) vapor pressure results but stated that their result were not conclusive and required further study.

It should be noted that these comparison tests were conducted at a V/L= 4, which at such a large expansion ratio reduces the sensitivity of the measurement to light ends. As you decrease the V/L to ratios more typically seen in storage and transport (V/L = 0.05 to 0.2) the sensitivity of the measurement to light ends is increased.

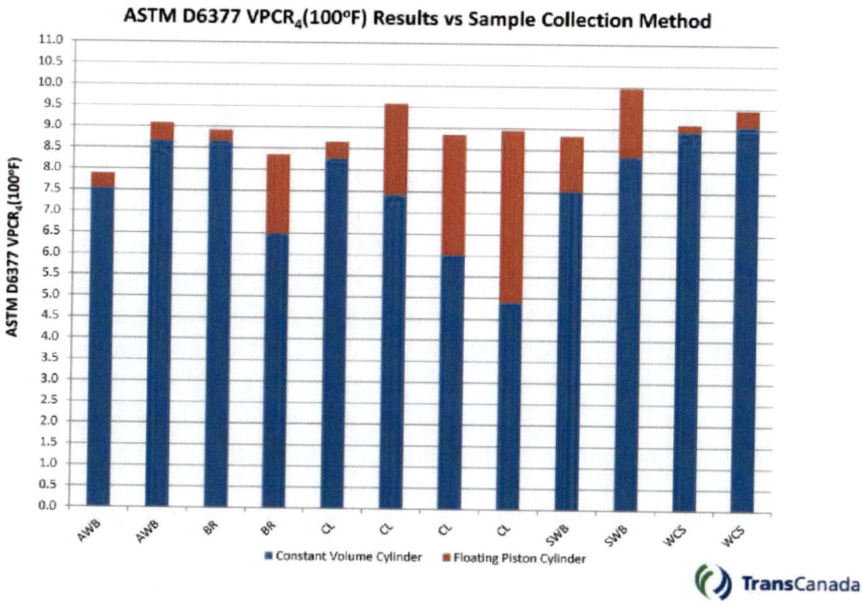

Figure 6-2. ASTM D6377 VPCR4 (100° F) VP Data of Crude Oil Sampled by FPC and Open Sampling Technique, Reproduced with Permission from Konecnik (2014).

6.2. Initial Boiling Point

The importance of accurately determining IBP of tight oils is illustrated by Table 6-1 (RP3000, 2014). IBP is a critical determinant in assigning the PG classification of a Class 3 flammable liquid.

Table 6-1. Criteria for Assignment of Class 3 Flammable Liquid PG[1]

Packing Group	Flash Point	Initial Boiling Point
I		≤35°C (95°F)
II	<23°C (73°F)	>35°C (95°F)
III	≥23°C, ≤60°C (≥73°F, ≤140°F)	>35°C (95°F)

[1]This table is for information purposes only and does not provide legal advice on compliance with regulations.

Table 6-2 (RP3000, 2014) is a list of IBP determination methods cited in either 49 Code of Federal Regulations (CFR) Subchapter C, Hazardous Materials Regulations or Canadian Transportation of Dangerous Goods Regulations (TDGR) as generally applicable to flammable liquid IBP determination for the purpose of PG assignment. Table 6-2 also includes ANSI/APIprovided comments regarding the applicability of each method to the highly specialized task of determining IBP for an oil with a significant presence of dissolved gases. Additional comments regarding each method (based on review of the latest available version of the method) are provided below.

6.2.1. HMR/TDGR-Cited IBP Determination Methods

ASTM D86-12 – This method was designed for analysis (via atmospheric pressure distillation) of distillate and spark ignition fuels, which rarely—if ever—contain appreciable quantities of dissolved gases (including methane, ethane, propane, and butanes [although butanes are added to winter gasoline blends]) that are often present in tight oils. Because the method specifies that IBP is determined to be the "...temperature reading at the instant the first drop of condensate falls from the lower end of the condenser tube," IBP values determined for tight oils that contain noncondensable gases are not likely to accurately represent the lowest temperature at which material is vaporized from the oils (ASTM, 2012).

ASTM D1078-11 – This method was designed for distillation of liquids that boil over a temperature range of 30-350°C. Because, like D86, IBP is determined based on recovery of condensed liquid, D1078 is not ideally suited for application to tight oils with appreciable quantities of dissolved gases.

ISO 3405:2011 – This International Organization for Standardization (ISO) method has been aligned with ASTM D86, and employs the same approach (recovery of condensed liquid) to establish IBP.

Table 6-2. IBP Determination Methods for Assignment of PG

Test Method	Applicable Range[2]	Result Type	Units	Comments
ASTM D86[1]	0°C to >250°C	Numeric	°C	For purposes of PG assignment using 35°C (95°F), a pass/fail using D86 can be applied. However, this test method may not be the most appropriate test method in measuring the IBP for crude oil with light components (i.e., methane to butanes).
ASTM D1078[1]	30°C to 300°C	Numeric	°C	Applicable for PG assignment purposes. This test method is not commonly used for wide boiling material, more for narrow range chemicals.
ISO 3405[1]	IBP=0°C, FBP <400°C	Numeric	°C	For purposes of PG assignment using 35°C (95°F), a pass/fail using ISO 3405 can be applied. However, this test method may not be the most accurate test method in measuring the IBP for crude oil with light components (i.e., methane to butanes).
ISO 3924[1]	>55°C (131°F)	Numeric	°C	Not applicable for IBP of less than 55°C (131°F).
ISO 4626[1]	-30°C to 100°C	Numeric	°C	Applicable for PG assignment purposes. This test method is not commonly used for wide boiling material, more for narrow range chemicals.

[1]Test methods listed in the *HMR* and the *TDGR*. This table is for informational purposes only and does not provide legal advice on compliance with the regulations.

[2]Applicable temperature ranges are as of the date of publication of this document.

Note: The use of alternative test methods to those listed in the *HMR* can be approved by U.S. DOT.

ISO 3924 – Based on its applicability to only materials with IBPs of greater than 55°C, this method was not reviewed.

ISO 4626 – 1980 (E) – This method was designed for determining boiling point distribution of volatile organic liquids such as hydrocarbons, esters, alcohols, ketones, ethers, and similar materials. Like the methods referenced above, ISO 4626 relies on recovery of condensed liquid as the means of establishing IBP.

6.2.2. Alternative Approach for IBP Determination

Because tight oils—especially un-weathered or "live" tight oils—typically contain significant levels of dissolved gases, the above-referenced methods are of limited value in establishing IBP with sufficient accuracy to enable appropriate tight oil PG classification. An alternative approach for tight oil IBP determination is described in ANSI/API RP3000 and summarized here. To minimize loss of volatile low-molecular-weight components, the crude tight oil should be sampled using a closed container (pressurized cylinder) method or other appropriate method for volatile liquids. Vapor pressure of the sample—at RVP measurement conditions of 100° F and vapor/liquid ratio of 4/1—is determined via ASTM D6377. If the measured RVP is equivalent to or less than 82.7 kPa (12 psi), IBP is determined via ASTM D7900, and if RVP is greater than 82.7 kPa, IBP is determined via either Gas Processors' Association (GPA) Standard 2103 or a version of ASTM D7900 that has been modified to include sample introduction techniques borrowed from GPA 2103. Table 6-3 lists key features of D7900 and GPA 2103.

Table 6-3. Alternative IBP Determination Methods for Assignment of PG

Test Method	Applicable Range[1]	Result Type	Units	Comments
ASTM D7900	Methane to n-nonane	Numeric	°C or °F	
GPA 2103	Methane to hexane	Numeric	Volume %	Should convert volume% to weight% for IBP calculation.

[1]Applicable temperature ranges are as of the date of publication of this document.

Note: The use of alternative test methods to those listed in the *HMR* can be approved by U.S. DOT

While ASTM D7169 (Standard Test Method for Boiling Point Distribution of Samples with Residues Such as Crude Oils and Atmospheric and Vacuum Residues by High Temperature Gas Chromatography) can be used for crude oil boiling point determination, this method is not recommended for application to tight crude oils that can contain significant quantities of dissolved gases and relatively light (C4-C8) hydrocarbons. Key reasons for non-recommendation are:

- The method uses capillary columns with thin films that result in incomplete separation of C4-C8 hydrocarbons in the presence of method-specified carbon disulfide solvent, which yields an unreliable boiling point distribution over this elution interval.
- The method typically results in co-elution of C1-C4 hydrocarbons with n-C5.

ASTM D7900 (Standard Test Method for Determination of Light Hydrocarbons in Stabilized Crude Oils by Gas Chromatography) was designed for calculation of boiling range distribution of hydrocarbons in stabilized crude oil up to and including n-nonane (C9). (To yield a full boiling point distribution for a crude oil, D7900 can be combined with D7169). For the purpose of D7900, stabilized crude oil is defined as having an RVP equivalent to or less than 82.7 kPa. IBP is defined as the temperature at which the first 0.5 weight% of the sample is eluted through the gas chromatography column, with temperature assigned as the actual boiling point of the pure compound (methane, ethane, propane, butane, isobutane, or a higher carbon number hydrocarbon) eluting at the point in the chromatogram that the initial 0.5 weight% value is reached. Although this method is able to provide valuable information regarding light end (C1- C9) composition of a crude oil sample, because the method is based on gas chromatography, it does not necessarily yield a definitively accurate IBP measurement. Because gas chromatography effects clean and complete separation of individual crude oil components, it is unable to account for or quantify the impact of azeotropic and other intermolecular interactions that are known to occur in complex mixtures like crude oils. While these intermolecular interactions are likely to have a relatively minimal impact on the temperature at which methane is released from a crude oil, they will exert increasingly more impact on the temperatures at which increasingly larger and decreasingly volatile species like ethane, propane, and butanes are released. This means that IBP values obtained via chromatography-based methods may not necessarily represent the actual temperature at which the first vapor emissions occur from a given crude oil in a setting outside of a gas chromatography column.

As stated above, for crudes with an RVP greater than 82.7 kPa, IBP can be determined via GPA 2103-3 (Tentative Method for the Analysis of Natural Gas Condensate Mixtures Containing Nitrogen and Carbon Dioxide) or a combined GPA 2103-D7900 method in which sample introduction into the chromatograph is performed via the procedure described in GPA-2103. The primary difference in these two 2103-based methods versus D7900 is the use of a liquid sampling valve capable of entrapping a fixed volume of sample at a pressure of at least 200 psi (1379 kPa) above the vapor pressure of the sample at valve temperature. Because IBP is again determined based on the boiling point of pure compounds as they elute from the chromatography column, neither method is able to account for or quantify azeotropic and other intermolecular interactions that could impact the actual temperature at which vapors are first emitted from a crude oil.

6.2.3. IBP As a Primary Determinant of PG

While the above summary is probably not inclusive of all crude oil-applicable IBP determination methods, the methods briefly described here are likely the most commonly used to determine IBP for the purpose of tight crude oil PG assignment. Without considering challenges associated with volatile crude oil sampling and handling prior to IBP determination, key limitations of these methods are:

- The use of condensation to establish the temperature at which vapors are first released from hydrocarbon mixtures that often contain significant levels of un-condensable materials.
- The use of pure compound concentrations and discrete boiling points to calculate IBP for complex hydrocarbon mixtures that are subject to intermolecular forces that impact the temperature and rate at which individual compounds are released as vapors.

6.3. Vapor Pressure

6.3.1. Functional Definition

The term vapor pressure for crude oils is confusing upon review of the literature. The term vapor pressure as used in the literature can mean True Vapor Pressure (TVP), BPP, RVP or now, with variable volume expansion vapor pressure analyzers, we can have vapor pressures determined at various expansion ratios.

6.3.1.1. True Vapor Pressure

The TVP of a pure substance is easily defined as the total pressure exerted by a gas in equilibrium with a liquid at a temperature of interest. The measurement of TVP for a pure substance is constant provided there are distinct vapor and liquid phases in the measurement cell.

The TVP of a mixture is still the total pressure of gas in equilibrium with a liquid, but the TVP of a mixture, unlike a pure substance, is also dependent upon the molar vapor and liquid volumes (V/L) at a temperature of interest. The addition of the V/L term is required due to the differing volatilities of the components in the mixture.

6.3.1.2. Bubble Point Pressure

The BPP of a pure substance or mixture is a special case of TVP where the V/L = 0. Also defined as the pressure at which the first incipient vapor bubble is formed at a temperature of interest.

6.3.1.3. Reid Vapor Pressure

Reid Vapor Pressure is a specialized method to determine the vapor pressure of an air saturated liquid at 37.8 C (100° F) at a V/L = 4. The RVP test method and history will be discussed below.

6.3.2. ASTM D6377 Standard Test Method for Determination of Vapor Pressure of Crude Oil: VPCR$_x$ (Expansion Method)

The ASTM D6377 test method provides a means to determine the vapor pressure of a crude oil at user defined temperature and V/L ratio. The method is commonly utilized with a V/L = 4 to approximate the original RVP measurements of ASTM D323. The D6377 method utilizes a highly automated analyzer comprised of a cylindrical measurement chamber with moveable piston to allow introduction of a pressurized liquid sample of a known volume into the measurement chamber. Once the chamber is isolated the piston is raised to the user defined V/L expansion point of interest. While the sample pressure is allowed to reach equilibrium the measurement chamber temperature is controlled to the user-defined temperature. The sample is determined to be at equilibrium when three successive pressure readings, taken at 30 second intervals, agree within 0.3 kPa (0.04 psi). The measured pressure at equilibrium is reported as the VPCR$_x$ (Tm, C or F). Where, x = V/L fraction, and Tm = test temperature, in °C or °F (ASTM, 2014d).

The use of D6377 complaint analyzers is becoming more widespread across the crude oil industry due to D6377's high repeatability and reproducibility. The sample conditioning requirements are minimal compared to D323 RVP and sample introduction can be accomplished in a pressurized system eliminating potential loss of highly volatile components in the sample. As noted in the NDPC study (Auers, et al., 2014) industry is utilizing the VPCR$_4$ (37.8° C, 100° F) as a replacement to the original D323 RVP measurement. To address industries desire to use the D6377 method as an alternate to the D323 method, the U.S. Environmental Protection Agency (EPA) on May 28, 2013, issued a letter to the API approving the use of the D6377 method for the determination of the vapor pressure of crude oils with vapor pressures which fall within the D6377 stated applicability of 25-180 kPa at 37.8° C (EPA, 2013). The EPA letter was silent as to the V/L expansion to utilize for the measurement, but industry has taken the intent to be measurement to be $VPCR_4$ (37.8° C).

6.3.3. ASTM D323 Standard Test Method for Vapor Pressure of Petroleum (Reid Method)

The ASTM D323 RVP method was originally developed from 1918-1930 by German chemist Reid, for the determination of natural gasoline vapor pressure. The method was accepted as ASTM Standard in 1930. The method is outlined below (ASTM, 2014e).

1) Sample acquired per ASTM D4057 into open container. Sample container to be filled to 70-80% full, balance is air.
2) Sample is chilled to (0-1° C), once container chilled it is opened to bring sample back to atmospheric pressure and shaken, air saturation step is repeated three times.
3) Vapor chamber is heated to 37.8° C, and remains air filled. Vapor chamber volume is four times the liquid chamber volume.
4) Chilled liquid sample in transferred to liquid chamber and attached to vapor chamber.
5) The total assembly is placed into 37.8° C water bath.
6) After five minutes, the assembly is removed from bath, shaken, returned to bath and pressure monitored. This is repeated five times every two minutes. Test is complete when two successive readings are the unchanged.

The test method takes care to saturate the sample with air at 0-1° C prior to testing. This is to remove any variations in the sample air saturation as received by the testing laboratory. It is the intent of the D323, as originally developed, "to report the partial pressure of the hydrocarbon, not including the contribution from dissolved air or gas." (Faulkner, 2000) The D4057 sampling method is specified in D323 which is an open sampling method. With the specified sampling method and the three air saturation steps in the sample preparation for testing, the D323 method presents the potential for the loss of light end components.

> The D323 method is limited to those materials with a RVP (at 100° F) less than 26 psia to avoid both vapor loss and errors from gas saturation, but this assumes that the lightest component in the sample is butane or butane propane mixtures. This is an artifact of the RVP method originally being developed for casinghead gasoline measurement in the mid 1920's in order to properly control propane content. (Faulkner, 2000)

The initial use of the method was for the vapor pressure testing of gasoline and the sample chilling and air saturation steps assumed that the sample contained only propane and heavier components. Therefore, in the original use of the method, the chilling to 0° C was sufficient to minimize potential sample losses. However, if the sample contains lighter components, methane, ethane, CO_2, N_2. H_2S these components can outgas during the air saturation steps. An example was run using the Peng Robinson equation of state to simulate the spiking of two 12 RVP oil samples, the first sample was spiked with butane to produce a 26 psi RVP at 100° F sample and the second sample was spiked with N_2 to produce the a 26 RVP at 100° F. When both samples are chilled to 0-1° C the butane spiked sample pressure is 7-8 psia, no vapor loss upon opening, and the N_2 spiked sample pressure is 17-18 psia, and will outgas hydrocarbons along with nitrogen upon opening the sample. (Faulkner, 2000)

Due to these recognized limitations to the D323 Reid method, the Canadian Crude Quality Technical Association (CCQTA) published a letter, stipulating that the D323 method is not appropriate for "live" crude oils, due to the loss of light ends from the sample (2014).

6.4. Overall Findings

6.4.1. The North Dakota Petroleum Council Study on Bakken Crude Properties

On August 4, 2014, the North Dakota Petroleum Council released a study on Bakken Crude Properties by the Bakken Crude Characterization Task Force. The scope of the study was "...to establish Bakken crude properties (Quality Characterization) and to understand if these properties pose transportation and handling risks unique to Bakken compared to other light crude oils" (NDPC, 2014), with specific interest in transport by rail. The report does not describe how crude oil properties relate to risk or how they may pose danger during transport. As such, it is not clear within this report what properties should be used to form a basis for comparison of transport risks with other crudes. The report does address a number of standard oil quality parameters that are readily available for Bakken and similar light sweet crudes that help to characterize the oils to some extent, though the value toward the specific issue of risk in rail transport is not established. The report asserts that "...the test results from this study are consistent with reported results from other sources...regarding the quality of Bakken" (NDPC, 2014) as it pertains to the comparison data presented from other source crude oils.

The NDPC report provides the largest compendium of Bakken crude oil data to date with

- API Gravity: 152 Samples
- D86 IBP: 152 Samples
- Vapor Pressure (D6377): 152 Samples
- Flash Point (D3278): 152 Samples
- Light Ends (IP344): 152 Samples
- Simulated Distillation (D7169): 111 Samples

A summary of the key analyzed data are presented below in Table 6-4:

Table 6-4. NDPC Summary of 152 Bakken Crude Sampling Data

Total 152 Samples	Avg	Min	Max
API Gravity	41.0	36.7	46.3
Vapor Pressure, psi	11.7	8.9	14.4
D86 IBP, 0F	99.5	91.9	106.8
Light Ends C2 -C4, LV%	5.45	3.33	9.30

The NDPC report supplies data which are utilized to determine DOT HMR Hazard Class, which is based upon vapor pressure and physical state (gas/liquid), and PG for flammable liquids based upon Flash Point and IBP. Flash point data per ASTM D3278, was reported as <73° F, which is lower criteria limit for flash point for HMR Packing Group determination for all samples. However, D3278 states "The test methods are applicable to paints, enamels, lacquers, varnishes, and related products having a flash point between 0 and 110° C (32 and 230° F) and viscosity lower than 150 St at 25° C (77° F)" (ASTM, 2011). More detailed reporting of the flash point data may provide further insight to the relative flammability of the Bakken versus other light crude oils.

The second measurement germane to PG determination is the IBP of the flammable liquid. The NDPC report IBP's for the 152 samples tested ranged from 91.9° F to 106.8° F. The PG I vs PG II determination threshold is 95° F. The report then goes on to explain the shortcomings of the D86 IBP method because "...D86 was not developed for wide boiling range materials like crude oil, with no specifically defined lab-operating parameters specified. Therefore, different labs used different operating conditions during testing, resulting in a wide variability of values for the IBP." (NDPC, 2014). The report then discusses the fact that the D86 test has the inability to truly measure the IBP of volatile liquids containing C_1-C_4 compounds due to these compounds low condensing temperatures at atmospheric pressure.

The NDPC report fails to report the actual measured flash point values and acknowledges the shortcomings of the D86 test method to accurately determine the IBP of C_1-C_4 containing liquids. Both measures, per DOT, bear directly upon the level of hazard of the liquids. This makes it difficult to then use these same two measures (flash point and IBP) to support the conclusion that Bakken oil flammability is commensurate with that of other light crude oils.

The NDPC report includes the vapor pressure [$VPCR_4$ (100° F)] for 152 samples from well sites, rail loading terminals, including five samples from the St James, LA receiving rail terminal. All samples were obtained in chilled one-quart glass jars and immediately sealed, chilled and transported to laboratory. As discussed in D323 RVP section, crude oils containing N_2, CO_2 and C_1-C_4 compounds can off gas at reduced sample temperatures when opened to the atmosphere. When the sample bottles are opened for sample withdrawal for analysis, light ends can off gas and thus reduce the measured vapor pressure. The NDPC reports all of its data at $VPCR_4$ (100 F) which is the industry standard as a legacy of the D323 method, and allows for ease of comparison with other oils. Utilizing the $VPCR_4$ (100° F) as basis to report vapor pressures is less relevant with respect to the transport environment, where tank ullages are typically 2-5%. For the purposes of determining actual transport tank pressure utilizing a V/L of 0.02 to 0.05 would more closely approximate tank car conditions. Utilizing the V/L of 0.02 to 0.05 increases the impact the dissolved air and gases have on actual "as received" tank car pressures, where vapor pressure measurements at V/L = 4 minimize the impacts of the dissolved air, gases and light ends.

The NDPC report states that sampling with FPCs in a closed sample was inconclusive versus bottle or open samples. They reported testing four parallel samples with at rail loading terminal to provide sample point with high enough pressure to allow sampling with FPC. No results were reported but the report stated that

> While some samples showed excellent agreement both with historic NDPC sampling and between the glass bottle and FPC samples at the pressurized sample point, others showed variation, with samples taken off the line having lower vapor pressure values than the samples collected from the tank. This implies that samples taken at the pressurized sample point downstream of the tank somehow lost light ends by comparison. (NDPC, 2014).

However as stated earlier, there has been data reported (Pichler and Lutz, 2014; Konecnik, 2014) that open bottle testing does provide lower $VPCR_4$ results versus those from FPC sampling.

The NDPC report does supply a vast quantity of well facility operating data and compares the impacts of well separation equipment (separator, heater treater) operating pressure and temperature upon measured vapor pressure and light ends content. The results of this analysis proved unremarkable and the recommendation of the report was to operate the separation equipment at as low of pressure and at as high of temperature as the system could safely permit. The NDPC study measurements were made only at the stock tanks downstream of the separation equipment, making it impossible to measure the true impact of the facility operating variables. No facility inlet crude oil data was collected to allow direct analysis of the impact of facility unit operating conditions on vapor pressure or light ends of the outlet crude oil.

The NDPC report does provide a breadth of data that greatly enhances the understanding of the Bakken crude oil properties and provides insight into the relative homogeneity of the oils from this producing horizon. The report does point out some of the shortcomings of the HMR testing as it pertains to crude oils in general and particularly to light crude oils which may contain C_1-C_4 compounds. The report further recommends that "based upon the findings of this study, the NDPC encourages all members to classify their BKN crude as a Class 3 PG I flammable liquid until a more definitive testing protocol is established." (NPPC, 2014)

6.4.2. Pipeline and Hazardous Materials Safety Administration Report

The PHMSA prepared a report as part of its overall effort called Operation Safe Delivery, which is "...examining the entire system of crude oil delivery." (PHMSA, 2014) The report issued in July 2014 provided the results of PHMSA sampling and analysis of 135 Bakken crude oil samples over a 10-month period from August 2013 to May 2014. The purpose of the PHMSA study was to insure that the Bakken crude oil was being properly classified per DOT HMR.

Table 6-5. Summary of PHMSA Sampling and Test Protocols

Sample Date	**Aug, 2013 – Nov, 2013**	**Feb, 2014 – May, 2014**
Samples Taken	47 Total Samples from rail loading facilities and cargo tanks, storage tanks, pipelines used to load rail cars and several were collected from cargo tanks. All samples were collected in accordance with ASTM 4057, "Standard Practice for Manual Sampling of Petroleum and Petroleum Products.	88 Total Samples from rail loading facilities and cargo tanks, storage tanks, pipelines used to load rail cars and several were collected from cargo tanks. Samples were collected via a syringe-style cylinder in accordance with ASTM 4057, "Standard Practice for Manual Sampling of Petroleum and Petroleum Products.
ASTM Tests Conducted	Standard Test Method for Vapor Pressure of Petroleum Products (Reid Method) (ASTM D323). Standard Test Method for Determination of Individual Components of Crude Oil (ASTM D6730 MOD). Standard Test Method for Water and Sediment in Crude Oil (ASTM D4007). Standard Test Method for Sulfur in Petroleum and Petroleum Products (ASTM D4294). Standard Test Method for Measurement of Hydrogen Sulfide in the Vapor Phase Above Residual Fuel Oils Hydrogen Sulfide Content (ASTM D5705). Standard Test Method for Density and Relative Density for Crude Oil (ASTM D5002). Standard Test Method for Flash Point by Tag Closed Cup Tester (ASTM D56). Standard Test Method for Distillation of Petroleum Products at Atmospheric Pressure (ASTM D86).	Standard Test Method for Vapor Pressure of Petroleum Products (Reid Method) (ASTM D323). Standard Test Method for Determination of Individual Components of Crude Oil (ASTM D6730 MOD). Standard Test Method for Measurement of Hydrogen Sulfide in the Vapor Phase Above Residual Fuel Oils Hydrogen Sulfide Content (ASTM D5705). Standard Test Method for Flash Point (FP) by Tag Closed Cup Tester (ASTM D56). Standard Test Method for Distillation of Petroleum Products at Atmospheric Pressure, Initial Boiling Point (IBP) (ASTM D86). Standard Test Method for Determination of Vapor Pressure of Crude Oil: VPCRx (Expansion Method) for both Vapor/Liquid ratios of 0.02 (at 122 °F) and 4 (at 100 °°F). U.N. Recommendations on the Transport of Dangerous Goods, Manual of Tests and Criteria, Chapter 37 (corrosion to aluminum and carbon steel).

The study reports that prior to the its launch the Federal Railroad Administration (FRA) “identified that most crude oil loading facilities were basing classification solely on a generic Safety Data Sheet (SDS), formerly known as Material Safety Data Sheets (MSDS)” (PHMSA, 2014). PHMSA noted that the SDSs supplied “were out-of-date with unverified data” (PHMSA, 2014). These findings prompted PHMSA to expand the scope to include sampling and analysis to provide more chemical property data to insure the oil was being properly classified. Table 6-5 summarizes the samples taken and ASTM test methods conducted.

It should be noted that the sample locations in the PHMSA report were from rail loading facilities, cargo tanks, pipelines used to fill rail cars, which are facilities downstream of the primary production facilities. The data reported in the PHMSA report was in line with the data presented in the NDPC 2014 report and American Fuel & Petrochemical Manufacturers (AFPM) 2014 report with the following exceptions.

The Flash Point data reported in the PHMSA reporting was performed per ASTM D56 and the reported range of flash point was <32° F to <73° F versus <73° F in the NDPC report and a range of -74.2-122° F in the AFPM report.

The IBP and the resultant PG classification for the PHMSA report yielded a much higher percentage of PG I samples versus that of the NDPC report. The average reported IBP from the PHMSA report is 87.0° F versus the 100.3° F average from the NDPC report and 69.6° F from the AFPM report. This is important given the PG I IBP temperature criteria is 95° F.

The PHMSA report does provide a new data point to the Bakken characterization database, $VPCR_{0.02}$(122° F). This is to estimate the vapor pressure of the oil at rail car conditions of roughly 2% ullage. The $VPCR_{0.02}$(122° F) readings ranged from 20.3 to 39.36 psia, with an average value of 27.98 psia. Although the rail car pressure data in the AFPM report did not indicate the bulk temperature of the reported refinery rail car pressures. And recognizing the previously stated shortcomings of the open sampling methods for vapor pressure determination, the PHMSA $VPCR_{0.02}$(122° F) readings are in the range as those reported in the AFPM report who reported an average of 8.5 psig (23.2 psia) and a maximum of 11.3 psig (26.0 psia). As stated earlier the $VPCR_{0.02}$ at varying temperatures will, with proper sampling methods, provide more sensitivity of the vapor pressure measurement than those measured at VPCR4.

Given that the PHMSA reports primary focus was on the verification of proper HMR Classification of the Bakken crude oil, as currently prescribed in regulation, the study was constrained somewhat to the regulatory test methods.

The exception being the $VPCR_{0.02}$(122° F) data collected. The stated sampling methods utilized were per D4057, which is an open sampling method. The study then states for the February-May 2014 sampling was performed with a "syringe-style" cylinder per D4057. It is unclear from this description what method was employed given no reference to a "syringe-style" cylinder is made in D4057.

6.4.3. American Fuel & Petrochemical Manufacturers' Report

The AFPM prepared a report, "A Survey of Bakken Crude Oil Characteristics Assembled for the U.S. Department of Transportation (DOT)," (Wybenga, 2014) to DOT in response to a query from PHMSA related to the characteristics of Bakken crude oil and their potential transportation risks. The report summarized information provided by 17 AFPM members, which included analyses of 1400 samples of Bakken crude oil that were acquired from various locations along the distribution chain between wellhead production points and refinery rail tank car receipt points. The report focused on six Bakken crude oil characteristics that are relevant to explosion hazards and government flammable liquid classification and transportation regulatory requirements. These characteristics, the number of data collected for each by the survey, and the number of different test methods included the following:

- Flash point: 77 samples based upon 3 different test methods
- Initial boiling point: 275 samples based upon 5 different test methods
- Vapor pressure (RVP): 807 samples
- Flammable gas content: 18 samples based upon 3 different test methods
- Hydrogen sulfide content: 37 liquid samples and 535 vapor samples
- Corrosivity to metal: 7 samples based upon 1 test method

Participation in the survey by AFPM members was voluntary. Sampling and test methodologies employed in acquiring the results were at the discretion of individual members. As a result, with the exception of corrosivity, multiple (i.e., two to five) standard test methods were employed by different members for each characteristic. The AFPM report provided no information regarding technique(s) used to collect and handle samples.

A summary of key survey property data are presented in Table 6-6.

Table 6-6. Summary of Bakken Crude AFPM Survey Data. (Wybenga, 2014)

Property	Number of Samples	Average	Minimum	Maximum
API Gravity	0	42 *		
Reid Vapor Pressure, psi	807	10.4 **	0.8	15.54
IBP, °F	275	84.2	36.0	152.4
Light Ends C_2-C_4, LV%	18	6.8 ***	2	11.9

* from a table of assay data for Bakken and non-Bakken crude oils (Wybenga, 2014)
** estimate based upon chart of "RVP Frequency for Bakken Crudes" (Wybenga, 2014)
*** estimate based upon chart of "C1 Thorugh C4 Frequency" (Wybenga, 2014)

In addition to the average vapor pressure value displayed in Table 6-6 that was calculated directly from data exhibited in the AFPM report, the report listed RVP values of 7.83 psia (from an assay) and 7.6 psia (as being "a normal RVP value"). No reference was provided for either value. (Wybenga, 2014) With regard to DOT's inquiry, the report concluded that, based upon member responses to DOT questions and test results, Bakken crude oil is in compliance with DOT HMR (49 CFR, Parts 105 to 180) and is properly categorized as a Class 3 Flammable Liquid. Furthermore, Bakken crude "does not pose risks significantly different than other crude oils or flammable liquids authorized for rail transport"; in fact, it "poses less risk than other flammable liquids authorized for rail transport in the same specification tank cars" (Wybenga, 2014). The report reached these conclusions by juxtaposing member information with four HMR classification criteria and requirements and a PHMSA question regarding the existence of flammable gas concentration thresholds which, if exceeded, prohibit rail transportation of Bakken crude oil (Wybenga, 2014):

1. HMR generally defines a liquid as substances with vapor pressures of not more than 300 kPa (43.5 psia) at 50° C and melting points at or below 20° C. RVP measurements were applied as a surrogate for vapor pressures, the survey observed a RVP range of 0.8 to 15.54 psia with a mean of roughly 10.4 psia, which is substantially within the range defined for a liquid; consequently, Bakken crude oil is classified as a Class 3 Flammable Liquid. The report observed that average values varied seasonally from 8 psia during warmer periods to 12.5 psia during colder periods.
2. HMR classifies flammable liquids into three Packing Groups based upon closed-cup flash points and initial boiling points as indicated in Table 6-7. Member data reported a range of flashpoints of -59° to 50° C (plus one

outlier of 110° C) and an initial boiling point range of 2.2° to 66.9° C with the majority at 35° ± 15° C. Based on these data and the criteria defined in Table 6-7, Bakken crude oil can be placed into Packing Groups I and II, with the exception of the outlier, which is in Group III (Wybenga, 2014). The report notes that at least some flash point values might be based on estimation methods and so might deviate from methods acceptable under HMR. Additionally, members reported issues with variability and repeatability of initial boiling point (IBP) measurements. This is significant because of the clustering of samples having IBP values near the dividing line between Groups I and II at 35° C.

Table 6-7. DOT Flammable Liquid Classification Criteria

Packing Group	Closed-Cup Flash Point	Initial Boiling Point
I		At most 35° C
II	Less than 23° C	More than 35° C
III	At least 23° C and at most 60° C	More than 35° C

3. HMR has established hazard communication requirements for materials that pose inhalation hazards—hydrogen sulfide is one such material. Consequently, members reported both liquid and vapor hydrogen sulfide concentration measurements that they had acquired. Of 405 measurements reported by one group of members of hydrogen sulfide in the vapor phase, only six were at concentrations greater than 2,000 ppm and almost 400 were less than 2 ppm. A separate member reported 129 vapor samples which exhibited a mean of 3,280 ppm and a maximum value of 23,000 ppm. All of the reported 37 liquid samples contained less than 10 ppm. The report's conclusion was that Bakken crude oil could present an inhalation hazard; thus for crude shipments with high hydrogen sulfide levels, HMR hazard communication requirements apply.

4. HMR classifies a substance as corrosive if it corrodes steel or aluminum at a rate of 6.25 mm (0.25 inches) a year. NACE International Test Method (NACE TM) 172 corrosivity test results that were received by AFPM had scores of B+ or B++ (B+ or better is required for pipeline transportation), implying HMR does not classify Bakken crude—by itself—as presenting a corrosivity hazard. However, contaminants that accompany Bakken crude, e.g., water, solids and hydrogen sulfide, might pose an issue.

5. The PHMSA inquiry included a request for information regarding the existence of limits of on flammable gas concentration for transportation. The

report replied by providing the ranges of flammable gas that were acquired by the survey. The survey received results from 18 samples that had been tested for flammable gas content. Results from six samples that were received from three members had a range of total flammable gas of 2%-3.5% and had the profile exhibited in the middle column of Table 6-8. A fourth respondent submitted the results of testing 12 samples which were distinctly higher than the other six samples with a range of total flammable gas of 5.9% to 11.9% and had the maximum concentration profile that appears in the right column of Table 6-8. In discussing flammable gas in Bakken crude oil, the report displays a table of light ends from assays of several crude oils in addition to Bakken. A portion of this table has been extracted and presented in Table 6-9. "Light ends" are defined similarly to "flammable gases" except they also include pentanes, C5. The report notes that the concentration of light ends in Bakken is similar to several and less than four of the non-Bakken crude oil assays. The report concludes that while the 12 sample results possess higher flammable gas content than the Bakken assay, such variation is common among individual samples and that the assay data provides a more consistent basis for evaluating light ends among crude oils.

Table 6-8. Flammable Gas Compositions (Liquid Volume %)

Compound	Six Sample Test Results	Twelve Sample Maximum Test Results
Methane (C1)	less than 0.01	0
Ethane (C2)	0.5	0.77
Propane (C3)	<2	3.4
Butanes (C4)	3-4	8.0
Maximum Total	3.5	11.9

Table 6-9. Properties of Several Light Crude Oils (Wybenga, 2014)

Crude Name	Origin	API	RVP (psia)	Liquid Volume % (constituents)
Arabian Super Light	Saudi Arabia	51	20.7	12.53 (C1–C4)
Eagle Ford	Texas	48	7.95	8.3 (C2–C5)
DJ Basin	Colorado	45	7.82	8.0 (C2–C5)
Saharan Blend	Algeria	43	7.46	8.1 (C2–C5)
Bakken	North Dakota	42	7.83	7.2 (C2–C5)

In addition to the above test results, the AFPM survey also reported rail tank car pressure measurements that were taken at the receipt point (i.e., refineries). The pressure profile possesses a shape similar to the RVP profile, with a mean value of 8.5 psig and a maximum of 11.3 psig. The report compared those values with the 35 psig minimum relief valve setting and the 240 psig minimum design burst pressure of older DOT 111 rail tank cars to demonstrate that an ample margin exists between Bakken crude characteristics and rail tank car design.

The AFPM report concluded, "This survey shows that Bakken crude oil does not pose risks that are significantly different than other crude oils and other flammable liquids authorized for transportation as flammable liquids." (Wybenga, 2014, 27) However, the report has fundamental limitations. First, it should be noted that, for data provided to AFPM that was acquired during routine testing to support manufacturing operations, the purpose of acquiring the data likely was more to identify unexpected crude oil characteristics or characteristics that might create processing problems in a load of crude oil than to acquire a scientific understanding of the material sampled. As a result, accuracy might not have been as important as consistency with past sampling and testing, especially when properties being measured are not as important to processing or are substantially within the operating parameters of the equipment that will process it. Second, the report provided little description of the sampling methodologies and qualifications of the persons or organizations collecting and analyzing the samples. Finally, since submission of test results was voluntary, sample collection was not systematic or comprehensive, which makes the representativeness of the data unknown. This makes the report somewhat anecdotal.

6.5. Data Comparisons

For the current discussion, the authors aggregated some of the currently available vapor pressure data pertaining to light crude oils and put these in context with SPR oils that have been studied in detail for vapor pressure properties. There is implied consensus across the literature that vapor pressure is pertinent to the discussion of safety around flammable liquids as it appears among the leading parameters typically evaluated in this context. Vapor pressure is one measurable property of a liquid that can be used to partially evaluate the volatility, though it is evaluated at equilibrium and does not give direct information on the rate of vapor formation or the amount of vapor formed at equilibrium. The specific relationship between vapor pressure and probability or severity of hazard with crude oils in transit has not been established, and that is a primary research issue to be addressed in possible future phases of this work. There are a number of challenges for comparing crude oil vapor pressure data. First, sampling methods are not consistent, and may have important shortcomings for retaining the lightest ends. The NDPC (Auers, et al., 2014) and PHMSA (2014) studies both contain numerous crude samples from the Bakken region, and most of those were obtained using an open capture method, storing the samples in sealed glass jars. As mentioned above in Section 6.4 two recent studies (Konecnik, 2014; Pichler and Lutz, 2014) indicate that open or unpressurized sampling can give low results for D6377 $VPCR_4$ relative to closed sampling. Their tests were run with crudes showing $VPCR_4$(100° F) in the range 7-10 psia, which is the same range as SPR oils, and a few psi lower than most of the NDPC and PHMSA Bakken oils. While "volatility" is sometimes cited as a property of interest for evaluating flammability risks, it is generally used as a qualitative term and must be quantified for purposes of comparison. Volatility of Liquefied Petroleum gases may be evaluated by ASTM D1837 (ASTM, 2011a), which reports the temperature at which a certain volume percentage of hydrocarbon liquid is vaporized as it is heated from its condensed liquid state. Its value toward use as a comparative property for flammability risks was not explored here. Note that crude oil properties can also change with time and exposure to ambient conditions whereby dissolved gases dissipate and the heavier components are left behind as the oil experiences a natural "weathering" process. As such, highly volatile oils can change quickly with time in a spill/release scenario.

6.5.1. Bases for Comparison

Comparisons in the following sections are drawn from data presented in the PHMSA (2014), NDPC (Auers, et al., 2014), and SPR sources. Parameters compared include API gravity, $VPCR_4$(100° F), BPP, gas-oil ratio (GOR), and light ends (C1-C4) compositions.

6.5.2. BPP and GOR Measurements at SPR

SPR routinely measures BPP, GOR, and flash gas compositions for its oils in storage in order to monitor oil quality and maintain its ability to deliver oil to market within performance specifications for stream BPP and GOR. The SPR received a pipeline delivery of Bakken oil in December 2012 that was analyzed for a suite of typical oil quality parameters as well as BPP and GOR. Selected data are reported below in Table 6-10, as well as in Figure 4-4 (fractional distillation summary). The SPR Bakken sample is referred to using the "SPR Bkn" acronym in figures and tables shown from here forward.

Table 6-10. Selected Measured Data for SPR Bkn Received at the SPR Bayou Choctaw Storage Site on Dec 12, 2012

Property	Value	Units
Relative Density, 60/60° F	0.8185	
API Gravity	41.4	
Sulfur	0.242	Mass %
Bubblepoint pressure @100°F	21.0	Psia
Gas-oil ratio @ 100°F, 14.7 psia	6.6	scf/stb

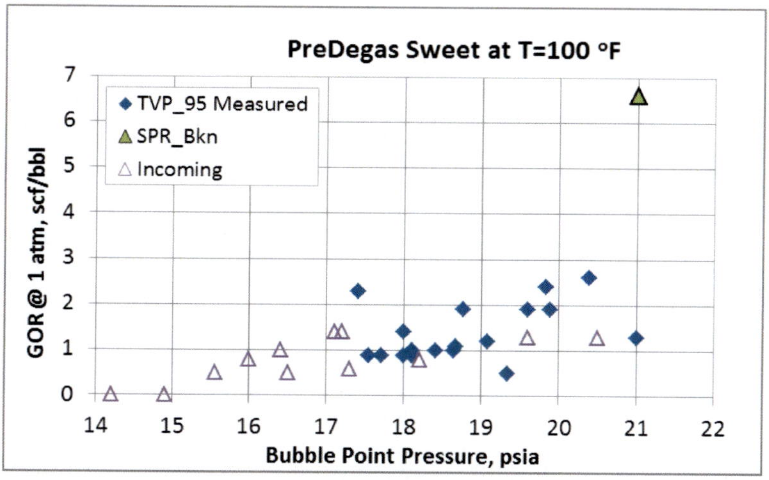

Figure 6-3. Plot of SPR Bkn Overlaid with Incoming SPR Crudes and also SPR Sweet Oils Currently in Storage (TVP_95 Measured).

The SPR Bkn may be put into context with other incoming crude streams received at SPR as well as sweet oils currently stored at SPR by comparing measured BPP and GOR, shown in Figure 6-3. Note SPR Bkn exhibits the highest BPP and GOR observed among the samples considered. Samples from incoming streams, including SPR Bkn, were obtained from pressurized flowing lines into a test separator. Samples from the SPR sweet inventory here represent ~50L spot samples collected from the mid-depth of ~10 million barrels SPR caverns using a floating piston-cylinder device operated on wireline. All SPR samples shown here were tested prior to any stabilization processing that may occur on-site. A general trend is seen among the data that

BPP and GOR increase together, which makes sense because both properties are related to the presence of light end gases in the whole oil.

6.5.3. Expansion Curves

GOR and BPP values represent two specific points on a pressure-volume expansion curve at a given temperature (T = 100°F in Figure 6-3), and plotting the entire curve is often instructive to better understanding the VLE behavior for an oil (recall Figure 4-5 and Figure 4-8).

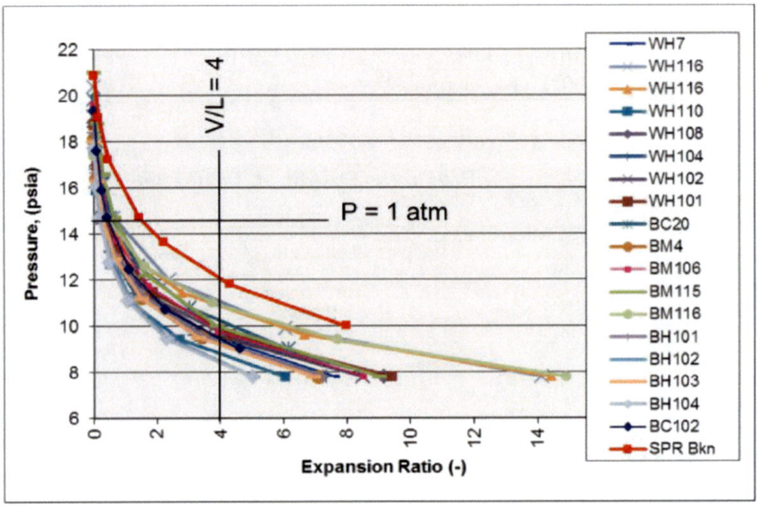

Figure 6-4. Plot of SPR Bkn Overlaid with SPR Sweet Oils Currently in Storage at SPR.

The test separator utilized at SPR (recall section 4.2.4) collects compositional data from the flash gas stream and is passed through an EOS model in order to build a whole oil compositional model for the feed stream. With an EOS model, it is possible to simulate pressure-expansion points outside of what was directly measured. A subset of the data presented in Figure 6-3 had sufficient quality data to run EOS expansions. These simulation results were plotted in Figure 6-4, showing vapor pressure versus expansion ratio for a number of SPR caverns (i.e. WH7) as well as the SPR Bkn sample. The figure is annotated to show P = 1 atm (which is the simulated GOR in V/L units), and V/L = 4 (which is the simulated equivalent of $VPCR_4$ @ T = 100° F). Note SPR Bkn is again the highest case for both pressure and V/L, but particularly as it departs from the rest of the data with V/L several times higher than simulated for the other SPR sweet oils. This illustrates an important concept in that oils exhibiting the same bubblepoint pressure (say 21 psi as the upper end shown here) may exhibit notably different GOR (or V/L). Understanding why this is requires a look at the underlying compositional data.

6.5.4. Flash Gas Compositions

An inspection of the flash gas compositions for SPR Bkn and WH108 reveals some key differences in the light ends, illustrated in Figure 6-5. First, SPR Bkn has much lower nitrogen than WH108, and this may be related to the transport mode. The SPR Bkn sample was transported to SPR across the country via pipeline. Many other SPR oils were moved to the U.S. through Gulf coast ports via tanker ships, which utilize an inert gas blanket, often nitrogen, to reduce fire and explosion hazards in the crude oil tanks. Nitrogen has a very high individual vapor pressure and can contribute to raising a mixture bubblepoint pressure, but has less effect on gas-oil ratio since it is in relatively low concentration in the feed stream. C1 (methane) mole % values are comparable between oils. Meaningful differences to the GOR appear when looking at C2-C4, where SPR Bkn exceeds WH108 several mole% for each component. While this is not necessarily striking in looking at the scale of the figure, operational experience at SPR combined with a body of similar data indicates that these seemingly small differences in the key C2-C4 components correlate with the higher GOR oils in the SPR inventory (see also Figure 6-11). Also note that the WH108 oil shown here compares well with typical SPR sweet oil composition and properties (data not shown here), so it stands as a reasonable representative of the sweet inventory at SPR.

6.5.5. Whole Oil Compositions

While flash gas properties are informative to understanding VLE behavior, crude oil is tested, stored, and transported as a liquid phase and the "whole" oil or feed stream composition will determine how it behaves in any process or transportation environment. The flash gas data presented in Figure 6-5 were passed through an EOS model and assumed to represent a saturated gas at separator P, T, flow conditions. From this, along with knowledge of average molecular weight and relative density of the saturated liquid, a whole oil model was built. The resulting compositional profiles for both oils are compared in Figure 6-6. The bar chart and the table present the same data, and the reader should reference the table where the scale on the bar chart is too small to show anything meaningful. Generally speaking, the SPR Bkn shows slightly higher mole fractions (~.01) for C2-C7, which is likely where the higher GOR is coming from.

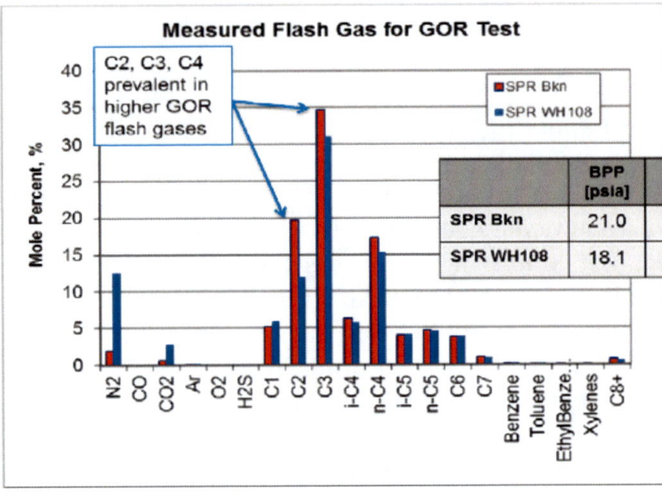

	BPP [psia]	GOR [scf/bbl]
SPR Bkn	21.0	6.6
SPR WH108	18.1	1.4

Figure 6-5. Comparison of Measured Flash Gas Compositions for SPR Bkn and WH108.

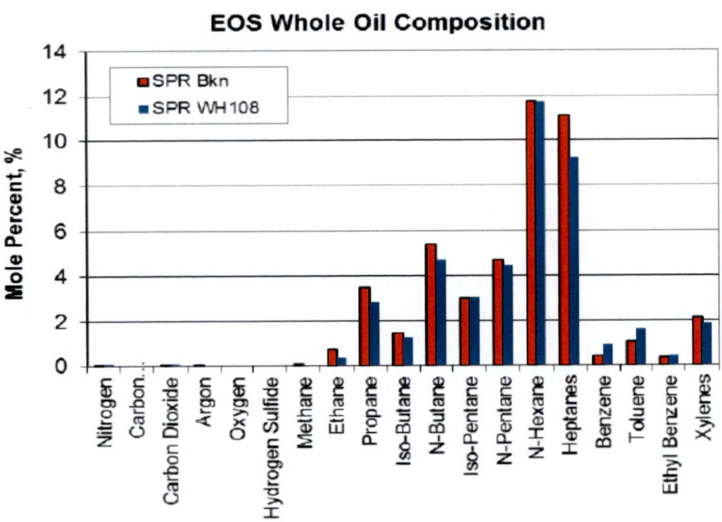

Figure 6-6. (Continued).

	SPR Bkn	WH108
	mole frac	mole frac
Nitrogen	0.0004	0.0006
Carbon Monoxid	0.0000	0.0000
Carbon Dioxide	0.0002	0.0007
Argon	0.0000	0.0000
Oxygen	0.0000	0.0000
Hydrogen Sulfid	0.0000	0.0000
Methane	0.0009	0.0004
Ethane	0.0073	0.0035
Propane	0.0345	0.0285
Iso-Butane	0.0145	0.0126
N-Butane	0.0541	0.0468
Iso-Pentane	0.0300	0.0301
N-Pentane	0.0468	0.0443
N-Hexane	0.1172	0.1173
Heptanes	0.1110	0.0927
Benzene	0.0044	0.0096
Toluene	0.0105	0.0162
Ethyl Benzene	0.0034	0.0042
Xylenes	0.0211	0.0183
Residual	0.5438	0.5745

Figure 6-6. Estimates of the Whole Oil Compositions for SPR Bkn and WH108 Determined by EOS Analysis of the Saturated Vapor Stream Data (see Figure 6-5) from a Test Separator.

Residuals in this analysis were treated as a single hypo-group, which has proven sufficient for simple BPP and GOR calculations in the SPR operating environment.

A review of the data so far indicates that the 2012 pipeline SPR Bkn receipt exhibits slightly higher BPP and notably higher GOR than most other receipts and current sweet inventory at SPR. Considering the composition, the flash gas from separator tests shows more C2-C4 for the SPR Bkn than a typical SPR sweet. The whole oil composition of SPR Bkn shows more C1-C7 than a typical SPR sweet. The margins on the compositional differences are small, however, and will not stand out upon inspection unless the implications and sensitivities are well-understood. The next step on the data comparison is to look at how SPR oils and the SPR Bkn compare with Bakken from field studies published by NDPC and PHMSA.

6.5.6. SPR Oils vs. Bakken Oils

While SPR has a large dataset on vapor pressure for its own crude oils, its value to the problem of characterizing tight oils is leveraged with comparison to the available literature data. Figure 6-7 overlays VPCR4(100° F) vs. API gravity for a range of SPR, NDPC, and PHMSA data. Note PHMSA did not list individual API gravity values, so the mean VPCR4 and 2 standard deviation (2σ) lines were computed and added to the figure. Scanning the figure from left to right, VPCR4 increases with API gravity as would be expected for incrementally lighter oils. The NDPC and PHMSA Bakken oils exhibit a consistent VPCR4 range from about 10-14 psia with a mean value around 12 psia. The SPR Bkn oil sits right in the middle of the Bakken oils on the plot, and would appear to be similar enough to say that it may exhibit properties typical of a Bakken oil for the purpose of initial comparisons.

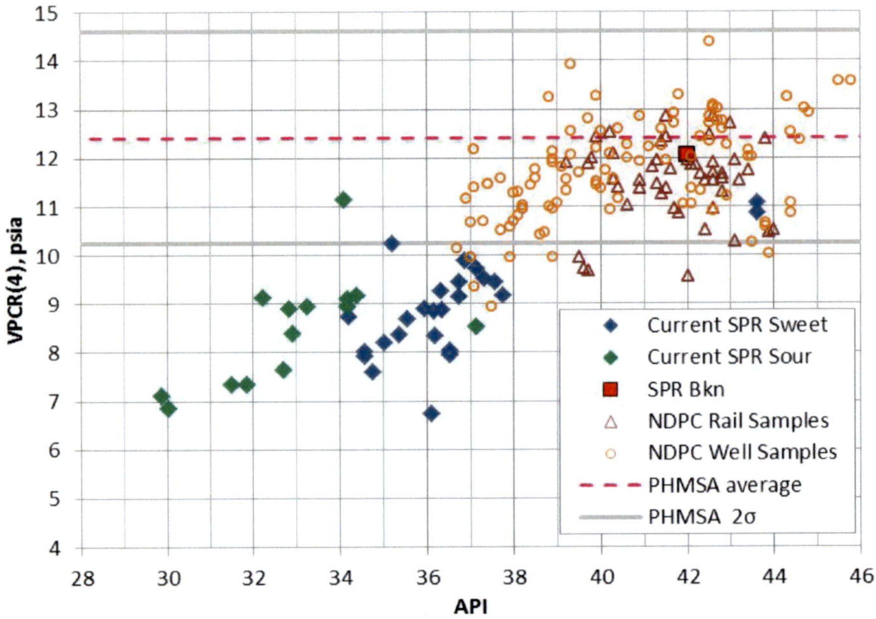

Figure 6-7. Overlay of SPR, PHMSA, and NDPC VPCR4(100°F) vs. API Gravity.

6.5.7. Light Ends Analysis

Some light ends data were available for comparison:

1) NDPC (Auers, et al., 2014) Appendix 8 – IP344 Light Ends data – Rail (vol%) for Bakken samples
2) PHMSA (PHMSA, 2014) table E Light Ends (liq vol %)
3) SPR Bkn and SPR WH108

Data were averaged and plotted in a bar chart in Figure 6-8. The same data are tabulated in Table 6-11. SPR Bkn was similar to the other Bakken oils in light ends volume, while SPR WH108 has slightly lower values than the other oils. Direct comparison between literature Bakken light ends and the SPR light ends is tenuous because the sampling and analysis methodologies were not identical. Within the SPR population, the SPR Bkn sample did show a higher light ends concentration via the same experimental methods as used across the SPR oils.

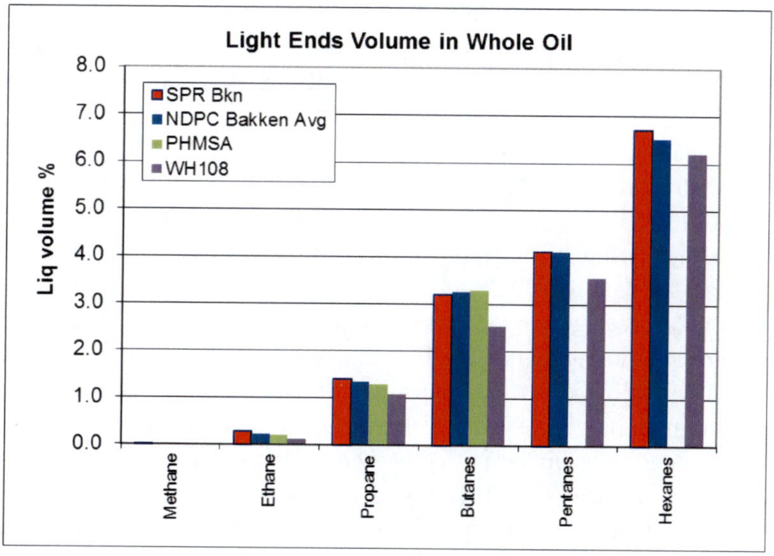

Figure 6-8. Average Light Ends Comparison.

Table 6-11. Average Light Ends Comparison

	NDPC	SPR Bkn	PHMSA	WH108
	liq vol %	liq vol %	liq vol %	liq vol %
Methane	0.00	0.02	0.00	0.01
Ethane	0.23	0.29	0.21	0.13
Prop a ne	1.36	1.40	1.30	1.08
Butanes	3.25	3.20	3.29	2.53
Pentanes	4.11	4.11	N/A	3.56
Hexanes	6.52	6.71	N/A	6.21

6.5.8. Vapor Pressure vs. Light Ends

The PHMSA VPCR data were plotted against the sum of C2+C3 liquid vol% in Figure 6-9. Two different VPCR conditions are given: $VPCR_4$(100° F), and $VPCR_{0.02}$(122° F). Note the $VPCR_{0.02}$ should return a reasonably close estimate to the bubblepoint pressure of the liquid at the given temperature 122° F. Also overlaid in red are the PHMSA samples collected with a closed syringe sampling technique. Recall all the rest of the PHMSA samples were collected with an open capture process. There are several observations worth discussing here. First, the increase in temperature from 100° to 122° F combined with the reduction in V/L from 4.0 to 0.02 induces significantly higher measured vapor pressures from an average around 12 psia up to around 28 psia. Vapor pressures are even higher for the syringe samples to as high as 39 psia.

There is also a trending apparent where increased C2+C3 vol% appears to correlate with increased VPCR. The correlation with C2+C3 vol% is stronger at $VPCR_4$(100° F) than at $VPCR_{0.02}$(122° F) with a tighter distribution. The behavior of the syringe samples is noteworthy in that they fall right on top of the open capture data for $VPCR_4$(100° F), but half of them segregate to values well above the average for open sampling at $VPCR_{0.02}$(122° F). This could be indicative of small amounts of the lightest gases (N_2, O_2, C1) that are present in small volumes but exert a high vapor pressure, and are likely lost in the open capture method. The effects of these gases tend to be most visible around bubblepoint conditions and are effectively diluted as more gases evolve at V/L = 4 condition. Konecnik (2014) noted dependence of $VPCR_4$(100° F) on sample capture method where open capture showed 0.25 to 4 psi lower values than closed capture. The PHMSA data show less dependence than Konecnik (2014) for their $VPCR_4$(100° F) data.

An even larger compendium of the $VPCR_4$(100° F) data versus C2+C3 vol% is shown in Figure 6-10. The data indicate an upward trending in $VPCR_4$(100° F) with C2+C3 vol% for all oils up to about 2% C2+C3, above which the relationship appears to break down, and $VPCR_4$(100° F) appears to limit out into band between 10-15 psia. The strong trending between 1-2 vol% makes sense, but the breakdown of the relationship above 2% is difficult to explain. There may be some value in exploring EOS predictions of behavior in this parameter space, but time and data availability did not allow for it in this report. The current authors would like to see if more compositional data are available for either of these datasets, with which EOS model predictions could be built.

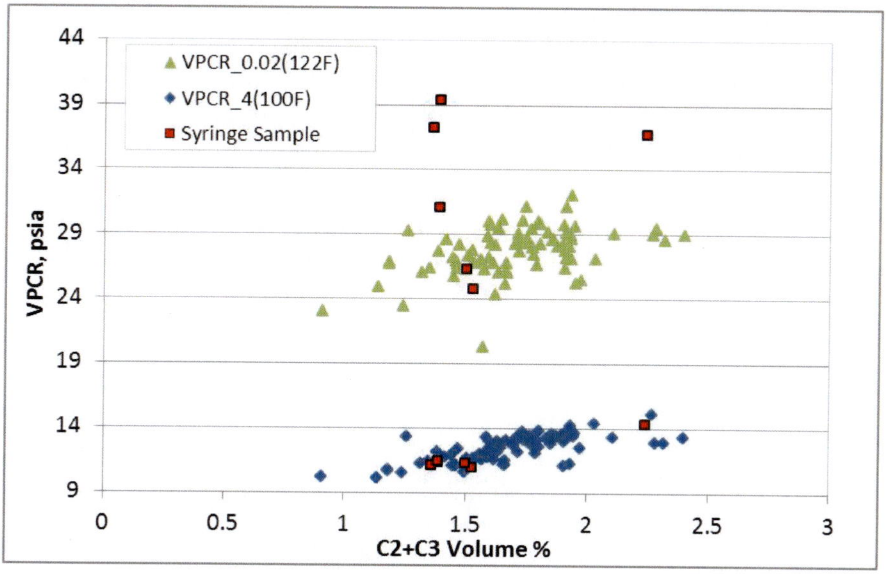

Figure 6-9. VPCR vs. C2 + C3 Volume % in Liquid, PHMSA (2014).

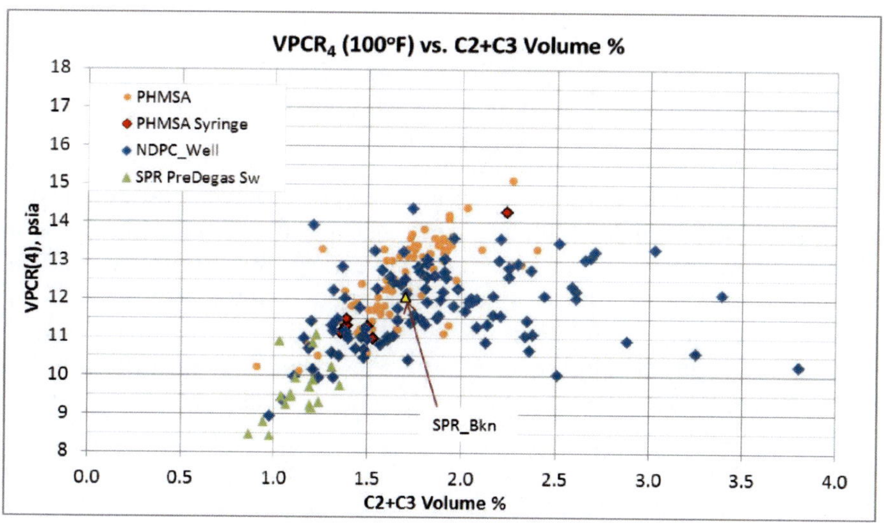

Figure 6-10. Overlay of VPCR4(100°F) vs. C2+C3 Liquid Volume % Data for PHMSA, NDPC, and SPR oils.

6.5.9. GOR vs Light Ends for SPR Oils

One last figure (Figure 6-11) is presented showing the measured GOR for SPR oils versus vol% C2 + C3. The GOR is not entirely governed by these selected components, but the data suggest that GOR may increase steeply with only modest fractions of volume % increases in C2 and C3.

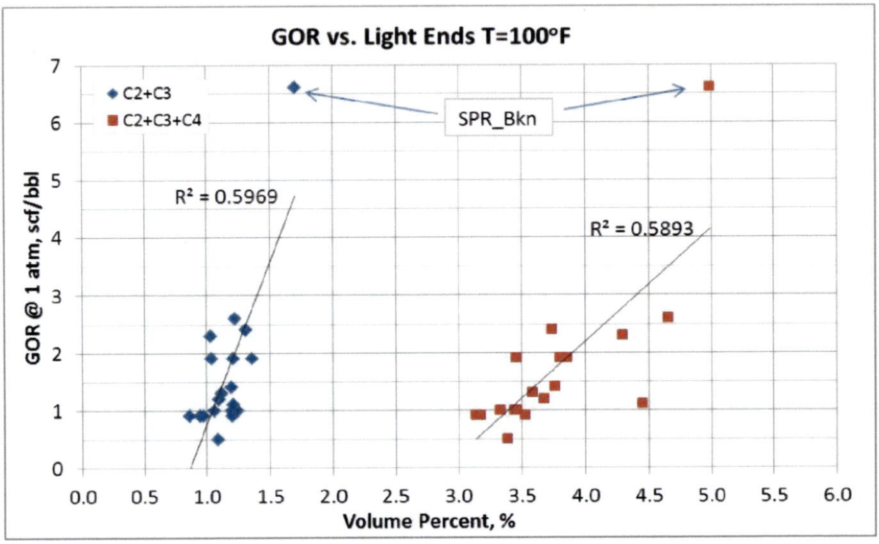

Figure 6-11. GOR vs. Light Ends Volume % in Whole Oil for SPR Oils.

6.5.10. Summary of Data Comparisons

In spite of some of the sampling and analysis methodology differences noted across the available literature data, VPCR4(100° F) appeared to be a useful starting point for comparisons, especially for Bakken oils. The distributions of VPCR4(100° F) values overlaid quite well for all Bakken oils and sources examined, with both open and closed sampling methods, up to a C2+C3 liquid volume of 2%. There is little data above 2% except for NDPC which showed peculiar behavior. There are published data (Konecnik, 2014; Pichler and Lutz, 2014) indicating that the open vs. closed sampling does impact VPCR4(100° F) values for crude oils in this vapor pressure range, though this was not observed for the small comparison set available in the PHMSA data.

Returning to the sampling method issue, the PHMSA testing at VPCR0.02(122° F) revealed some conspicuous differences depending on open vs. closed sampling, which should be investigated further. Careful compositional analyses for these samples, to include sensitive measures of light ends mole% coupled with EOS modeling would be useful in evaluating why the $VPCR_{0.02}$(122° F) values diverge.

Oils received and stored at the SPR are comparatively heavier and exhibit lower BPP and GOR than the Bakken oils from all data sources explored here. Most notable was that the Bakken pipeline sample received at SPR in December 2012 exhibited GOR that was ~5x higher than the average sweet oil in current SPR inventory. A combination of compositional analysis and EOS modeling indicate that the likely cause for this difference is a slightly higher whole oil mole% C2-C7 in the SPR Bkn oil than typical at SPR. To the extent possible with available data, the SPR Bkn sample appeared to fit well within the expected property ranges for $VPCR_4$(100° F) and light end compositions when compared with the NDPC and PHMSA Bakken datasets.

A summary look at average $VPCR_4$(100° F) for the oils considered in this literature survey is given in Table 6-12. Bakken oils, independent of source, appear to exhibit vapor pressures in the range of about 10-14 psia, with mean values 11.5-12.5 psia. SPR sweet oils exhibit $VPCR_4$(100° F) from about 7-11 psia, with a mean value of 8.9 psia.

Table 6-12. Average $VPCR_4$(100° F) and Associated 2σ for Crude Oil Datasets

Source	VPCR4(100°F) [psia]	2*σ [psia]
NDPC Bakken (rail)	11.5	1.6
PHMSA Bakken	12.4	2.2
SPR Bkn	12.0	-
SPR Sw	8.9	2.0

7. Potential Combustion Events in Rail Transport

The objective of this section is to describe combustion events and their associated hazards that could result from a rail car accident. In addition to the individual hazards associated with these events, the cascading potential for ignition of nearby infrastructure should be considered in a hazard analysis. Smoke propagation should also be considered since it too can be a hazard, such as when it is drafted through a building's ventilation system. Human behavior during evacuations also needs to be considered since it affects the number of casualties/injuries and, hence, total damage. With large populations and complicated pathways for exit strategies, effective evacuation efforts can be extremely difficult. All of these factors should be considered to understand the damage potential. The various combustion events are provided in Figure 7-1 in the form of an event tree. The rupture of a tank can occur either by a Boiling Liquid Expanding Vapor Explosion (BLEVE) or by mechanical or thermal damage that does not cause a BLEVE. The following sections will describe these combustion events and their associated hazards resulting from both of these types of tank ruptures.

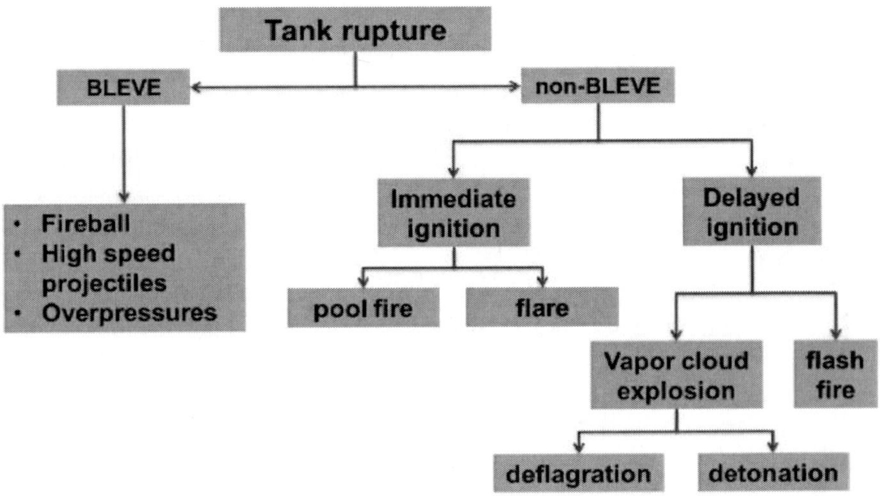

Figure 7-1. Combustion Event Tree from a Rail Car Tank Rupture.

7.1. Bleve

A rail tank car can rupture by means of a boiling liquid expanding vapor explosion (BLEVE), which is commonly defined as "an explosion resulting from the failure of a vessel containing a liquid at a temperature significantly above its boiling point at normal atmospheric pressure" (CCPS, 1999). The sudden release of the liquid that is above its boiling point can occur due to failure by various causes such as uncontrolled pressure build-up, wall-thinning due to hoop stress and high-temperature material degradation, severe bending stress due to thermal stresses induced by temperature variation on tank top and bottom, and mechanical damage due to impact, corrosion, or poor manufacture

(Birk, 1995; Abbasi and Abbasi, 2007). This type of explosion is considered a mechanical rather than chemical explosion. In a severe accident it is common to have some rail cars rupture from a crash where the fuel spills and ignites to form a pool fire that can engulf adjacent rail cars. As an engulfed rail car is heated, the fuel vaporizes within the car, which increases the pressure to a level that can exceed the pressure relief capacity and ultimately the design pressure of the container. Once the tank ruptures, the liquid fuel will achieve a superheated state due to the lower atmospheric pressure and will undergo a rapid phase change expansion. This expansion will result in vapor generation and atomization of liquid fuel that will simultaneously ignite to form a spherical partially pre-mixed flame known as a fireball. The combination of atomization and enhanced turbulent mixing allows the fuel to be combusted in a very short time. Additionally, there will be a shock wave generated from the BLEVE. The tank can fragment into pieces from the BLEVE or it can fail in other modes depending on the scenario. Fragments are not a necessary outcome from a BLEVE. If the tank car is not constrained by neighboring tank cars or other infrastructure, then the most likely failure mode will be the generation of several fragments that form projectiles that travel at high speeds and can be thrown large distances (~ 1 km). These projectiles will provide the greatest range of hazard impact as compared to the fireball or shock wave. There can be, however, scenarios where the tank cars can be sandwiched or piled-up together. The neighboring surfaces of adjacent tank cars can then provide a constraint to direct a BLEVE-driven failure to surfaces of least resistance and surfaces with the most thermal weakening, such as those enclosing the vapor space. The released fuel can then undergo a rapid vapor explosion and fireball. Evidence of this type of scenario is provided in the following three excerpts taken from a report by the Transportation Safety Board of Canada on investigating the accident that occurred in Lac-Mégantic, Quebec, Canada (2014). The first excerpt indicates that a large fire and multiple explosions occurred, the second excerpt provides a description of a thermal tear, and the last excerpt indicates that four cars experienced thermal tears with no fragments (TSBC, 2014).

> Page 2:
>
> *"At about 0115, the train derailed near the centre of town, releasing about 6 million litres of petroleum crude oil, which resulted in a large fire and multiple explosions."*
>
> Page 46:
>
> *"A thermal tear occurs when a tank car is exposed to elevated temperatures such as that from a post-derailment fire. As the*

temperature inside the tank rises, the product vapourizes, causing an increase in both its internal pressure and the stresses in the tank wall. If the pressure is not relieved, the tank ruptures. Ruptures involving the sudden release of built-up pressure can result in large explosions and fire."

Page 46:

"Examination of the 63 derailed tank cars showed the following:

- *None of the cars were equipped with thermal protection.*
- *Four cars that had sustained only minor impact damage due to the derailment experienced thermal tears, resulting in an energetic release.*
- *The length of the thermal tears ranged from 1.6 m to 4.4 m. No fragments of tank material were separated as a result of the thermal tears.*
- *All of the thermal tears were situated in the vapour space, and the PRDs were located in the liquid space.*
- *The car with the largest thermal tear (4.4 m) (Photo 27) was equipped with a PRD with an STD pressure of 75 psi, whereas the car with the smallest thermal tear (1.6 m) had a 165-psi PRD.*
- *Two tank cars experienced a thermal tear within approximately 20 minutes after the fire began."*

Source: Transportation Safety Board of Canada (2014).

Figure 7-2. Post-accident Site at Lac-Mégantic, Quebec.

Figure 7-2 shows the post-accident site taken from the report, which indicates the constraining configuration caused by the abutted cars; Figure 7-3 shows a rail car ruptured by a thermal tear. Based on observations of video coverage there are indications of BLEVEs occurring since large fireballs were accompanied by explosions. Also, the outward rupture pattern shown in Figure 7-3 also supports evidence of a pressure wave expansion indicating that a BLEVE occurred.

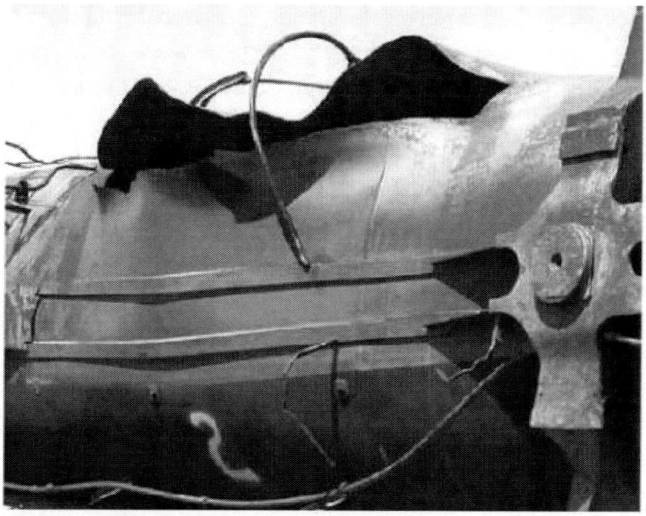

Source: Transportation Safety Board of Canada (2014).

Figure 7-3. Car with Thermal Tear.

7.1.1. Missiles and Blast Overpressure

Abbasi and Abbasi provided an overview of BLEVE mechanisms, accidents, and hazard analysis methods and discussed missile fragments and overpressure modeling (2007). The prediction of missile fragments such as number, velocity, mass, range, direction, and impact potential is dependent on the container and fuel under consideration. This is also true of the blast overpressures from vapor expansion.

Most research on missiles and overpressures have been performed on LPG vessels (Abbasi and Abbasi, 2007). Experimental testing is needed to understand the missile fragment behavior and overpressures produced from the BLEVE of a rail car container that transports tight oil.

7.1.2. Fireballs

Fireballs generated from BLEVEs are partially pre-mixed diffusion flames, which rapidly combust due to enhanced turbulent mixing and atomization from the vapor explosion resulting from a BLEVE. Experiments producing the largest fireballs were collected by Dorofeev, et al. (1995) in which experiments were performed using gasoline, kerosene, and diesel fuel of amounts from 0.1 to 100 tonnes. Fireballs were generated with fuel-rich vapor clouds and high explosives involving deflagrations and detonations, which are defined in Section 7.2.3. Data on the maximum fireball radius, duration, and total emitted energy for the different fuel masses was collected (Dorofeev, et al., 1995). The height of the fireball from center to ground varies from 3/4 to 1.5 times the fireball maximum diameter. The radiative heat flux for diesel fuel and kerosene is around 200 kW/m^2 (Dorofeev, et al., 1995). The aspect to note from the data is that there is not a wide variation in fireball radius, duration, and radiated energy among the fuels for large releases. Pertaining to the current application of interest, the data indicated that a 100 tonne release, approximately equivalent to a 30,000 gallon rail tank car assuming a fuel density of 800 kg/m^3, would produce a fireball diameter of around 200 m and duration of about 10-20 seconds. Empirical correlations that determine the length and time scales of a fireball are of the form $L = aM^b$ where a and b are empirical constants, M is the mass of fuel, and L represents radius, height, or duration. Thus, the amount of fuel involved is the most significant factor contributing to the severity of the fireball.

7.2. Non-BLEVE Outcomes

A non-BLEVE outcome is a scenario in which a rupture in the tank releases fuel without a vapor explosion. The outcome of the combustion event will depend on whether there is immediate ignition or delayed ignition. A pool fire or a flare can result with immediate ignition, whereas a vapor cloud explosion or a flash fire can result with delayed ignition. Each of these combustion events are discussed in the following sections.

7.2.1. Pool Fires

If a tank ruptures below its liquid level the fuel will spill and, if immediately ignited, will form a pool fire. Pool fires are typically defined as the burning of liquid fuel pools and are classified as highly buoyant diffusion flames. Diffusion flames are defined as the burning of fuel not initially mixed with an oxidant as in pre-mixed flames. The behavior of pool fires is highly dependent on the environment such as topography, wind, shape of the pool, and whether the fire is in an enclosure. The pertinent parameters for hazard evaluation are burn rate, flame height, and surface emissive power. The surface emissive power (SEP) is the energy emitted per unit time per unit area.

Wind can increase the burn rate, tilt the flame, and create fire whirls on the downwind side of the flame. Fire whirls tend to shed smoke to expose the luminous portions of the flame. Fire whirls can also be created with various topography and enclosure configurations. The burn rate affects flame height; higher burn rates result in an increase in flame height, which increases the view factor. View factors are used to determine how much radiative flux an object receives. Thus, an increase in flame height and, hence, view factor will result in an increase in the heat flux to an object, thereby increasing the thermal hazard distances.

Pool shape also affects the flame height. For instance, consider the configuration of a trench fire where the height is proportional to the shortest dimension. A very narrow trench fire provides a much lower view factor than a circular pool of an equivalent area. The difference could be an order of magnitude depending on the trench dimensions.

Water can affect hazard distances by changing the production and oxidation of soot, typically decreasing production and increasing oxidation. For optically-thin flames, where gas-band radiation dominates, the suppression of soot will reduce the net heat flux radiated to a distant object, but for optically-thick fires in which the flame is soot saturated, the effect of water vapor will be to suppress smoke formation with the consequences of increasing the net heat flux radiated to an object outside of the fire.

In optically thin pool fires, the local volume of flames is not able to absorb the local emitted radiation before it leaves the flame envelope, so the flame appears transparent. Pool fires will be optically thin up to a certain diameter (~1 m) and then transition to becoming optically thick where the flame is no longer transparent and the local soot production becomes saturated to the point that local radiation emission is absorbed within the flame envelope. The radiation emitted on the outer surface of the flame envelope originates at a layer near the surface. The SEP is then a function of the surface area of the flame and not the volume as for optically thin fires. Thus, the SEP will increase to reach a maximum at a pool diameter that is fuel dependent, typically around 2-3 m for many hydrocarbons. As the scale of the fire increases past this maximum, the rate of soot production overcomes the rate of soot oxidation and smoke begins to form. Eventually a diameter is reached in which so much smoke is formed that it nearly covers the entire flame except for a region near the pool surface where a continuous luminous zone persists. Smoke is a result of incomplete combustion, which is affected by radiative losses and limited oxygen supply and is made up of a mixture of gases, vapors, and particulate matter from a fire. Carbon particulates, or soot, is included as a particulate matter of smoke and is responsible for the luminosity of the fire as well as the black clouds often seen around fires. A sufficient layer of black smoke will absorb a significant portion of the radiation, resulting in a much lower emission to the surroundings and, hence, reduce thermal hazard distances. There are, however, luminous zones that appear periodically through the smoke layer due to the dynamic puffing nature of pool fires. Also, as noted previously, wind can create downwind fire whirls that can cause smoke shedding. Thus, when calculating hazard zones, these factors should be taken into account.

With increasing pool diameter, the average SEP over the flame will first increase with increasing fire diameter due to reaching the optically thick limit, and then decrease with further increases in fire diameter due to the increasing smoke layer coverage. Pool fire experiments performed by Koseki and Mulholland using Arabian light crude oil, heptane, toluene, and kerosene demonstrated this behavior (1991). They performed experiments on circular pool fires up to 2 meters in diameter and a square pool of 2.7 meters, and found that the radiation emitted decreases with increasing pool diameter, which reflects the increase in smoke production. Thus, with tight oils it is anticipated that similar behavior would occur since this is observed for many hydrocarbons.

For an object engulfed in the fire there is no smoke blockage effect, and heat flux to the object can vary from 100 kW/m^2 to 400 kW/m^2 depending on the environment. Based on several experiments conducted at Sandia, typical heat flux values of around 150 kW/m^2 are measured for JP-8 under quiescent conditions (Suo-Anttila and Gritzo, 2001a and 2001b). However, in wind conditions heat fluxes of up to 400 kW/m^2 can be attained on the downwind surfaces of engulfed objects (Gritzo, et al., 1997; Blanchat and Manning, 2001; Suo-Anttila and Gritzo, 2009). This is relevant to engulfed rail cars where the Code of Federal Regulations, 49 CFR PART 179: "Specifications for tank cars" requires that plates of similar tank material be subjected to a pool fire for 100 minutes and torch fire for 30 minutes (49 CFR). For the pool fire test the plate must be 10/16" thick, no less than 1 ft by 1 ft, and subjected to a temperature of 871° C to simulate a pool fire. Assuming black body radiation, this would correspond to about 97 kW/m^2. Note that the rail cars that carry tight oil are currently 7/16” thick rather than 10/16”. It is recommended that, to evaluate performance, experiments be performed with a higher temperature or heat flux level and with the actual thickness of the rail cars.

Koseki and Mulholland also measured burn rate as a function of pan diameter (1991). The data indicates that the burn rate asymptotes to a value of approximately 4 mm/min. It is anticipated that tight oils would have a burn rate similar to this value since many hydrocarbons asymptote to a burn rate between 3-6 mm/min under quiescent conditions, but as noted above can increase under wind conditions.

7.2.2. Flares

The rupture of a tank above the liquid level will release vapors and, if immediately ignited, can form a flare the size of which is determined by the size of the rupture. The thermal damage from flares to people is of main concern in an operational industrial setting (Brzustowski, 1976), but typically other combustion events that occur during an accident, such as a pool fire, will cause a much larger thermal hazard footprint. A flare has the potential to cause thermal damage to adjacent tank cars, initiate a BLEVE of adjacent tanks, and serve as an ignition source. Note that a flare will reduce pressures within a tank by burning off vapors, thereby facilitating the prevention of a BLEVE; thus, flares have intentionally been created by emergency personnel as a mitigation approach.

7.2.3. Vapor Cloud Explosions

If vapors released from a tank rupture are not immediately ignited a vapor cloud can develop and propagate. If it reaches an ignition source there is potential for a vapor cloud explosion, which is a rapid release of energy. Vapor cloud explosions from the combustion of flammable fuel/air mixtures are classified as either a detonation or deflagration. In a deflagration, the mixture burns relatively slowly, on the order of 1 to 10s of meters per second for hydrocarbon/air mixtures, though speeds will greatly exceed this if the flame is accelerated due to flow interaction with obstacles. This flame speed is in contrast to a detonation where the flame front travels as a shock wave followed closely by a combustion wave, which releases energy that sustains the shock wave. The detonation velocity for hydrocarbon/air mixtures is on the order of 2000-3000 m/s. Consequently, detonations generate very high overpressures, and, hence, are more damaging than deflagrations. Deflagrations can produce overpressures of up to around 0.8 MPa (115 psi) with obstacles and under 10 kPa (1.5 psi) without obstacles, while detonations produce overpressures up to around 2.0 MPa (300 psi). Open air vapor cloud explosions using low energy ignition sources will result in a deflagration without damaging overpressures (Leyer, et al., 1993). Additional hazards from explosions are fragments resulting from damaged structures. The amount of explosion overpressure is determined by flame speed, which is a function of the turbulence created within the vapor cloud and the amount of vapor within the flammability limits. There is a distinction between explosibility and flammability limits, but usually the flammability limit range is slightly broader than the explosibility range; thus, more conservative range is used.

The type of explosion that occurs depends on the strength and geometry of the ignition source, fuel, degree of confinement, and obstacle density. A deflagration may transition to a detonation when there is a high degree of confinement or blockage such as with closely spaced obstacles (Leyer, et al., 1993). The obstacles will create turbulence, which enhances the burning surface area and increases the local burning velocity. Detonations are more easily achieved with fuels of a smaller detonation cell width, a high-energy ignition source such as high explosives, and a high degree of confinement. Most vapor cloud explosions will result in a deflagration, which still can cause significant damage since most structures will be severely damaged with overpressures of 3-10 psi. Even with lower pressures, significant damage can still occur depending on the duration of the overpressure and characteristics of the structure.

7.2.4. Flash Fire

There are scenarios in which delayed ignition of a vapor cloud does not result in an explosion; that is, there is not a rapid release of energy and, hence, there are no damaging overpressures. The burning vapor cloud is then termed a flash fire, which poses thermal and asphyxiation hazards, the extent of which depends on the size of the vapor cloud. In a flash fire the flame will start in the region of ignition and then propagate through the cloud at a speed that is affected by wind conditions. If the wind direction is counter to the direction of flame propagation, the flame speed can be decreased to the extent that it is intermittently halted. The burning can also be heterogeneous with regards to flame height and soot production, that is, there may be times and regions where a tall flame occurs, and other times where the flame is approximately the height of the vapor cloud and a combination of transparent and luminous flames occur.

7.3. Thermal Impacts

Among the aforementioned combustion events, fragments generated from a BLEVE have the potential to provide the largest hazard footprint, at least initially following an accident. Fire propagation to nearby infrastructure has the potential to exceed the fragment range. It is difficult to rank the remaining combustion events since their severity will depend on various factors. The severity of a pool fire will depend on the size of the pool and the proximity of surrounding infrastructure. The thermal impact from a fireball will depend on the amount of fuel involved. And, the hazard footprint from a vapor cloud will depend on the amount of dispersed vapors and whether an explosion occurs. A thermal hazard evaluation involves calculating heat flux levels as a function of distance away from the fire. The impact of various heat flux levels is provided in Table 7-1. Often, the distance at which radiation from the fire has attenuated to a heat flux level of 5 kW/m^2 is used within industrial settings to determine the thermal hazard exclusion region. It is assumed that workers will have more protective clothing and places to take shelter. In settings for the general public, such as playgrounds, the criteria may be lowered to a heat flux level of 1.6 kW/m^2.

Table 7-1. Injury/damage of Various Heat Flux Levels and Exposure Times

Heat Flux (kW/m^2)	Injury/Damage
1	Typical value on a sunny day
1.6	Can be tolerated for long durations
5	Will cause pain in 15 – 20 seconds and second-degree burns after 30 seconds
37.5	Fire-resistant structures suffer damage after short duration. Metal fatigue after short-tomedium exposure (~10 min)

8. Combustion Properties for Hazard Calculations

Ignition is very difficult to predict since it is a function of chemical kinetics, heat, and mass transfer. Chemical kinetics involves having knowledge of all chemical reactions involved in combustion of a fuel and their reaction rates as a function of composition and temperature. Knowledge of detailed chemical mechanisms is limited for complex fuel mixtures (Dagaut and Cathonnet, 2006). More information exists on reaction mechanisms for simple fuels, but typically involves many species and reactions. For example, the detailed chemical mechanism for ethanol involves 56 species and 351 reactions (Marinov, 1999). In addition, heat transfer plays a major role in determining whether ignition will occur and if the burning is sustained. Mass transfer is also important since insufficient mixing of fuel and oxidant will not allow burning. Heat and mass transfer are highly affected by turbulence, which, in conjunction with chemistry, spans a vast time and length scale. Thus, it is not currently feasible to predict ignition using simulation based on first principles since solving all of the scales involved would be computationally prohibitive. Thus, the approach to classifying fuels with regards to ignition is empirically based. The following section first discusses parameters indicating a fuel's propensity to ignite and then discusses combustion properties required to perform a hazard calculation regarding the combustion events discussed in Section 6.

8.1. Flammability

There is no single parameter that defines the degree of flammability of a fuel, rather several parameters are relevant. They include:

> **Flashpoint:** Temperature that results in a vapor concentration in air corresponding to the lower flammability limit. When this temperature is reached there will be a flash of flame without sustained burning. The fire point is the temperature at which sustained burning occurs and is higher than the flashpoint.
>
> **Flammability limits:** Range of vapor concentration in the air that will support combustion. These are termed lower flammability limit (LFL) and upper flammability limit (UFL).
>
> **Auto-ignition temperature:** Minimum temperature at which a fuel-air mixture ignites.
>
> **Minimum ignition energy:** Minimum energy required to ignite a flammable fuel-air mixture.
>
> **Burning velocity:** Velocity at which a fuel-air mixture issuing from a burner burns back to the burner.

A fuel with a lower flashpoint, wider range of flammability limits, lower auto-ignition temperature, lower minimum ignition energy, and higher maximum burning velocity is considered more flammable. From a fuel classification and regulation standpoint, the main parameter considered relevant is the flashpoint. For operational handling, the above parameters are useful. However, it important to stress that in an accident scenario enough energy will be generated to cause ignition, far exceeding any hydrocarbon flammability classification threshold. Thus, the probability is extremely high that a combustion event will occur for a severe accident and additional parameters are needed in order to understand the hazards.

8.2. Properties for Hazard Calculations

To evaluate the thermal hazards from the combustion events discussed in Section 7, there are various parameters that must be obtained, some at small scale and some at large scale. The properties and relevant combustion event and test scale is provided in Table 8-1. The details of experimentally obtaining these properties will be provided in a subsequent report that discusses test plans.

Table 8-1. Combustion Events, Associated Property Required for Hazard Analysis, and Test Scale

Combustion Event	Properties	Scale
Applicable to all events	• Heat of combustion • Flammability limits • Boiling point temperatures of components • Density • Molecular weight • Composition in liquid/gas phases	Small; can be conducted in the laboratory.
Pool fire	• Burn rate • Surface emissive power • Flame height	~1 to 10 m Bund and free spill
Fireball/BLEVE	• Geometry • Surface emissive power • Duration • Fragment characterization (velocities, geometry, range) • Overpressures	Rail car
Vapor Cloud (flash fire, explosion)	• Gas composition	Rail car or Reduced-scale rail car

REFERENCES

46 CFR. (2014). Chapters 81, 121, and 551.

49 CFR, PHMSA, Hazardous Materials: Enhanced Tank Car Standards and Operational Controls for High-Hazard Flammable Trains, 49 CFR Parts 171, 172, 173, 174, and 179. Retrieved from http://www.phmsa.dot.gov /pv_obj_cache/pv_obj_id_9F3CE4CB7A6FCBD43CE1757CD2E 41F85A4C41000/filename/Signed_Proposed_Rulemaking-for_High-Hazard_Flammable_Trains.pdf

Abbasi, T., and Abbasi, S. (2007). The boiling liquid expanding vapour explosion (BLEVE): Mechanism, consequence assessment, management. *Journal of Hazardous Materials, 141*, 489-519.

Accurate, Inc. (n.d.-a). *Vertical and Horizontal Separators*. Retrieved December 12, 2014, from http://www.accuratecoatings.com/vertical seperators.htm

Accurate, Inc. (n.d.-b). *Freewater Knockouts*. Retrieved December 12, 2014, from http://www.accuratecoatings.com/freewater.htm

ANSI/API. (2014). *Classifying and Loading of Crude Oil into Rail Tank Cars*. Recommended Practice 3000. Washington, DC: American Petroleum Institute.

API. (2011). *Crude Oil Category Assessment Document*. High Production Volume Testing Group. Washington, DC: American Petroleum Institute.

API. (2014). *Manual of Petroleum Measurement Standards*. Chapter 8.4 Standard Practice for Sampling and Handling of Fuels for Volatility Measurement. Washington, DC: American Petroleum Institute.

Association of American Railroads (AAR). (2014). Moving Crude Oil by Rail. *AAR*. September 2014, p. 4.

ASTM International. (2010). Standard Test Method for Flash Point by Tag Closed Cup Tester. *ASTM D56-05*. West Conshohocken, PA: ASTM International.

ASTM International. (2011). Standard Test Method for Flash Point of Liquids by Small Scale Closed-Cup Apparatus. *ASTM D3278-96*. West Conshohocken, PA: ASTM International. ASTM International (2011a). Standard Test Method for Volatility of Liquefied Petroleum (LP) Gases[1]. *ASTM D1837-11*. West Conshohocken, PA; ASTM International.

ASTM International. (2012). Standard Test Method for Distillation of Petroleum Products at Atmospheric Pressure. *ASTM D86-12*. West Conshohocken, PA: ASTM International.

ASTM International. (2013a). Standard Test Method for Determination of Light Hydrocarbons in Stabilized Crude Oils by Gas Chromatography. *ASTM D7900-13E1*. West Conshohocken, PA: ASTM International.

ASTM International. (2013b). Standard Test Method for Flash Point by Pensky-Martens Closed Cup Tester. *ASTM D93-13*. West Conshohocken, PA: ASTM International.

ASTM International. (2014a). Standard Practice for Manual Sampling of Petroleum and Petroleum Products. *ASTM D4057-12*. West Conshohocken, PA: ASTM International.

ASTM International. (2014b). Standard Practice for Obtaining LPG Samples Using a Floating Piston Cylinder. *ASTM D3700-14*. West Conshohocken, PA: ASTM International. ASTM International. (2014c). Standard Practice for Sampling and Handling of Fuels for Volatility Measurement. *ASTM D5842-14*. West Conshohocken, PA: ASTM International.

ASTM International. (2014d). Standard Test Method for Determination of Vapor Pressure of Crude Oil: $VPCR_x$(Expansion Method). *ASTM D6377-14*. West Conshohocken, PA: ASTM International.

ASTM International. (2014e). Standard Test Method for Vapor Pressure of Petroleum Products (Reid Method). *ASTM D323-08*. West Conshohocken, PA: ASTM International.

Auers, J. R., R. M. Couture and D. L. Sutton. (2014). The North Dakota Petroleum Council Study on Bakken Crude Properties. *Bakken Crude Characterization Task Force*. Bismarck, ND: North Dakota Petroleum Council.

Birk, A. (1995). Scale effects with fire exposure of pressure-liquefied gas tanks. *J. Loss Prev. Process Ind., 8*, 275-290.

Blanchat, T., and Manning, L. (2001). *Mock B52 Bomb Bay Fire Experiment Data and Analysis for Model Validation and Development.* SAND2002-0145. Albuquerque, NM: Sandia National Laboratories.

Brzustowski, T. (1976). Flaring in the Energy Industry. *Prog. Energy Combst. Sci., 2*, pp. 129- 141.

Canadian Crude Quality Technical Association. (2014). *CCQTA Information Regarding the Measurement and Reporting of Light Ends and Vapor Pressure of Live Crude oil.* CCQTA. Retrieved from http://www.ccqta.com/files/CCQTA%20Information%20Regarding%20the%20Measurement%20and%20Reporting%20of%20Light%20Ends%20and%20Vapor%20Pressure%20of%20 Live%20Crude%20oil.pdf

Cattaneo, Claudia. (2014). Oil-by-rail shipments from Western Canada to triple in next two years amid pipeline crunch. *Financial Post*. Retrieved from June 9, 2014, http://business.financialpost.com/2014/06/09/canadas-oil-industry-cuts-long-termproduction-growth-forecast-to-4-8m-bpd/?__lsa=9a7a-44d3

Center for Chemical Process Safety (CCPS). (1999). *Guidelines for Consequence Analysis of Chemical Releases.* New York: American Institute of Chemical Engineers.

Continental Resources, Inc. (2014). *Bakken Central Tank Battery Design.* September 23, 2014 testimony before the North Dakota Industrial Commission, Case Number 23084 and Order Number 25417.

Dagaut, P., and Cathonnet, M. (2006). The Ignition, Oxidation, and Combustion of Kerosene: A Review of Experimental and Kinetic Modeling. *Prog. Energy Comb. Sci., 32*, 48-92.

Diverse Energy Systems. (2013). *Separation: Vertical, Horizontal & Skidded Heater Treaters.* Retrieved December 11, 2014, from http://des-co.com/wp-content/themes/radial/uploads/2013/09/ DES_Line-Cards_08-Separation.pdf

Dorofeev, S., Sidorov, V., Efimenko, A., Kochurko, A., Kuznetsov, M., Chaivanov, B., et al. (1995). Fireballs from Deflagration and Detonation of Heterogeneous Fuel-rich Clouds. *Fire Safety Journal, 25*, 323-336.

Environmental Protection Agency (EPA). (2013). (ALT-101) Approval of ASTM Method to Determine Vapor Pressure of Crude Oil, memo to P Watkins-API, May 28, 2013, Washington D.C.: U.S. EPA. Retrieved from http://epa.gov/ttn/emc/approalt/ALT101.pdf

Esser, Charles. (2014). Rail vs. pipelines: how to move oil: a growing shift in the way North America ships oil. *IEA Energy: The Journal of the International Energy Agency.* retrieved May 2, 2014, from http://www.iea.org/ieaenergy/issue6/rail-vs-pipelines-how-to-move-oil.html

Faulkner, R. J. (2000). *CCQTA update on recent activities on VP test measurement on crude oils.* Calgary, Alberta: CCQTA Conference. Retrieved from http://www.ccqta.com/files/TVP%20of%20Crude%20Oil % 20 Activities%20- %20Nov%202000.pdf

Feller, M. and H. J. Mcdonald. (1950). Determination of Liquid-Vapor Equilibria - New Dew Point Bubble Point Apparatus. *Analytical Chemistry 22(2)*: 338-340.

Frittelli, J. et al. (2014a). *CRS Report R43390, U.S. Rail Transportation of Crude Oil: Background and Issues for Congress.* December 2014.

Frittelli, J. et al. (2014b). *CRS Report R43653, Shipping U.S. Crude Oil by Water: Vessel Flag Requirements and Safety Issues.* July 2014.

Gardner Denver Nash. (2008). *Crude Stabilization Systems*. Retrieved December 13, 2014, from gdnash.com.br/ing/pdf/1055%20Crude%20Stailization%20systems.pdf

Gas Processing Association Canada (GPAC). (2014). *Strengthening the Testing and Classification Framework for Crude Oil by Rail*. GPAC Testing and Classification Working Group. Gas Processing Association Canada.

Gritzo, L., Moya, J., and Murray, D. (1997). *Fire Characterization and Object Thermal Response for a Series of Tests with a Large Flat Plate Adjacent to a Large JP4 Fuel Fire*. SAND97-0047. Albuquerque, NM: Sandia National Laboratories.

Herr, M. (2014). Standard is Set. *Petroleum News Bakken* 3(35): 13.

Konecnik, C. (2014). *HPLIS/Gas Chromatograph vs ASTM 6377 VPCR Results*. TransCanada. Crude Oil Quality Association Conference, San Francisco, CA, October 30, 2014. http://www.coqa-inc.org/docs/default-source/october-2014--san-francisco--joint-coqaccqta/10312014--konecnik--update-on-tvp-project.pdf?sfvrsn=2

Koottungal, Leena. (2014). 2014 Worldwide Refining Survey. Oil & Gas Journal 112(12). Koseki, H., and Mulholland, G. (1991). The Effect of Diameter on the Burning of Crude Oil Pool Fires. *Fire Technology*, 54-65.

Kringstad, Justin J. (2014). *ND Oil and Gas Research Council Update*. ND Pipeline Authority, presentation to the ND Oil and Gas Research Council, Bismarck, ND. September 15, 2014.

Kylling, Øyvind Widerøe. (2009). *Optimizing Separator Pressure in a Multistage Crude Oil Production Plant*. Norwegian University of Science and Technology, June 2009. Retrieved October 29, 2014, from http://www.divaportal.org/smash/get/diva2:347760/FULLTEXT01.pdf

Leyer, J., Desbordes, D., Saint-Cloud, J., and Lannoy, A. (1993). Unconfined Deflagrative Explosion without Turbulence: Experiment and Model. *J. Hazardous Materials, 34*, 123-150.

Lord, D. L., R. Allen and L. Eldredge. (2007). *Summary Report on October 2007 Grabner Temperature Tests at Degas Plant*. Geotechnology & Engineering Department. Albuquerque, NM: Sandia National Laboratories. U.S. Strategic Petroleum Reserve.

Lord, D. L. and D. K. Rudeen. (2010). *Strategic Petroleum Reserve Crude Oil Equation of State Model Development--Current Performance Against Measured Data*. SAND2010-4526. Geotechnology & Engineering Department, Albuquerque, NM: Sandia National Laboratories.

Mansoori, G. A. (2009). A unified perspective on the phase behaviour of petroleum fluids. *International Journal of Oil Gas and Coal Technology 2(2)*: 141-167.

Marinov, N. (1999). A Detailed Chemical Kinetic Model for High Temperature Ethanol Oxidation. *Int. J. Chem. Kinetics, 31*, 183-220.

Moritis, Guntis. (2001). Special report: future of EOR and IOR. *Oil & Gas Journal*. Retrieved December 12, 2014, from http://www.kindermorgan.com/business/co2/article_052001.pdf

Murray, D. (2014). "Canadian Crude Oil Quality--Handle with Care!" *presented at* Canadian Heavy Oil Association Meeting, Edmonton, Alberta, Canada, 27-Jan-2014, Canadian Crude Oil Quality Technical Association. Retrieved from: http://www.ccqta.com/files/CHOA%20 Presentation%20Jan%2027%202014%20V4%20r2.p df

North Dakota Industrial Commission. (2014a). *Order No. 24665*. Retrieved from https://www.dmr.nd.gov/oilgas/or24665.pdf

North Dakota Industrial Commission. (2014b). *Working Draft Case Number 23084 Order Number 25417*. November 13, 2014.

North Dakota Industrial Commission. (2014c). *Case Number 23084 Order Number 25417*. December 9, 2014.

North Dakota Industrial Commission. (2014d). *Crude Oil Processes*. August 28, 2014. Retrieved December 11, 2014, from http://www.dmr.nd.gov /oilgas/ presentations/NDIC082614CrudeOilProcesses.pdf

North Dakota Industrial Commission (2014e). *Activity Update NCSL Bismarck*, June 11, 2014. Retrieved March 11, 2015, from https://www.dmr.nd.gov/oilgas/presentations/ActivityUpdate2014-06-11NCSLBismarck.pdf

Oasis Petroleum. (2014a). Oasis Facility Operations, September 23, 2014 testimony before the North Dakota Industrial Commission, Case Number 23084 and Order Number 25417.

Oasis Petroleum. (2014b). *Investor Presentation*. Retrieved December 5, 2014, from http://www.oasispetroleum.com/wp-content/uploads/2014/11/2014-11-OAS-IRPresentationvFINAL.pdf

Oil and Gas Mineral Services. (2011). *Mineralwise® Free Water Knockout: definition*. Retrieved December 15, 2014, from http://www.mineralweb.com/library/oil-and-gas-terms/free-waterknockout-definition/

Pipeline and Hazardous Materials Safety Administration (PHMSA). (2014). *Operation Safe Delivery Update*. Washington, D.C: U.S. Department of Transportation.

Pichler, H. and J. Lutz (2014). Why Crude Oil Vapor Pressure Should be Tested Prior to Rail Transport. *Advances in Petroleum Exploration and Development 7(2)*: 58-63. Retrieved from http://www.cscanada.net/index.php/aped/article/view/5098, DOI: http://dx.doi.org/10.3968/5098

Recommended Practice (RP) 3000. (September 2014). *Classifying and Loading of Crude Oil into Rail Tank Cars*. ANSI/API RP 3000, First Edition.

Schlumberger, Ltd. (2014). *Oilfield Glossary: free-water knockout*. Retrieved December 15, 2014, from http://glossary.oilfield.slb.com/en/Terms/f/free-water_knockout.aspx

Suo-Anttila, J., and Gritzo, L. (2001a). *Thermal Measurements for a JP-8 Pool Fire in Cross-Flow with a C-141 Fuselage Section Located on the Leeward Edge*. SAND2001-0313. Albuquerque, NM: Sandia National Laboratories.

Suo-Anttila, J., and Gritzo, L. (2001b). *Thermal Measurements from a Series of Tests with a Large Cylindrical Calorimeter on the Leeward Edge of a JP-8 Pool Fire in Cross-Flow*. SAND2001-1986. Albuquerque, NM: Sandia National Laboratories.

Suo-Anttila, J., and Gritzo, L. (2009). The Effects of Wind on Fire Environements Containing Large Engulfed Cylinders. *Comb. Sci. Tech., 181*, 68-77.

Thermal-Fluids Central. (2011). *Representative Pure Alkane Phase Diagram*. Thermalfluidscentral.org. Global Digital Central. Retrieved from https://www.thermalfluidscentral.org/

Transportation Safety Board of Canada. (2014). *Runaway and main-track derailment, Montreal, Maine & Antlantic Railway Freight Train MMA-002 Mile 0.23, Sherbroooke Subdivision, Lac-Mégantic, Quebec, 06 July 2013*. Transportation Safety Board of Canada: Railway Investigation report R13D0054.

U.S. Department of Energy (DOE). (2014). *Quadrennial Energy Review Task Force Secretariat and Energy Policy and Systems Analysis Public Stakeholder Meeting: Infrastructure Constraints in the Bakken*. DOE. August 8, 2014.

U.S. Department of Transportation (DOT). (2014). *Pipeline and Hazardous Materials Safety Administration Safety Alert*, January 2, 2014.

U.S. Energy Information Administration (EIA). (2014a). Retrieved from http://www.eia.gov/tools/glossary/index.cfm?id=T

U.S. Energy Information Administration (EIA). (2014b). EIA projections show hydro growth limited by economics not resources. *Today in Energy*, June 10, 2014.

U.S. National Transportation Safety Board (NTSB). (2014). Tank car numbers from presentations by panel on tank car safety (panel 1). *NTSB forum, Rail Safety: Transportation of Crude Oil and Ethanol*. April 22–23, 2014.

Valerus. (2012). *Eagle Ford Complete Well Pad Solution*. Retrieved December 3, 2014, from http://www.valerus.com/project-profiles/united-states/eagle-ford-shale-texas/

Valerus. (2013a). *Eagle Ford Wellhead Solutions*. Retrieved December 3, 2014, from http://www.valerus.com/resources Solutions_DataSheets1.pdf Valerus. (2013b). *Rockies Oil Shale Play Wellhead Solutions*. Retrieved December 3, 2014, from http://www.valerus.com/resources/Rockies-Oil-Shale-Play-Data-Sheet1.pdf

Valerus. (2013c). *Eagle Ford Central Gathering Facility*. Retrieved December 3, 2014, from http://www.valerus.com/resources Gas_Gathering_DataSheets1.pdf Valerus. (2013d). *Well Pad Solutions: Marcellus/Utica Shale Plays*. Retrieved December 3, 2014, from http://www.valerus.com/resources/24374_Valerus_Marcellus_Utica_Regional_WellPad_flye r_PROD_without_crops1.pdf

Valerus. (2014a). *Eagle Ford: 2,500 BPD (17 m^3/hr) Condensate Stabilizer*. Retrieved December 14, 2014, from http://www.valerus.com/project-profiles/united-states/eagle-fordshale-texas-2/

Valerus. (2014b). *Utica: Condensate Stabilizers, 0140000*. Retrieved December 14, 2014, from http://www.valerus.com/ project-profiles/united-states/utica-condensate-stabilizers

Whiting Petroleum (2014). *Whiting Heater Treater Experience*. September 23, 2014 testimony before the North Dakota Industrial Commission, Case Number 23084 and Order Number 25417.

Wybenga, F. (2014). *A Survey of Bakken Crude Oil Characteristics Assembled for the U.S. Department of Transportation*. Dangerous Goods Transport Consulting Inc., American Fuel & Petrochemical Manufacturers.

GLOSSARY

API gravity: Common oil industry unit of measure for liquid density of a crude oil. See equation 4-1 in text for mathematical definition.

Auto-ignition temperature: Minimum temperature at which a fuel-air mixture ignites.

Boiling liquid expanding vapor explosion (BLEVE): An explosion resulting from the failure of a vessel containing a liquid at a temperature significantly above its boiling point at normal atmospheric pressure.

Bubble point pressure (BPP): The pressure at which gas bubbles first form within a liquid at a known temperature. This is also a special case of the true vapor pressure where the volume ratio of gas to liquid is zero (V/L = 0), and any incremental increase in temperature or decrease in pressure will incur gas formation and V/L > 0. The definition implies that the liquid is contained at a high pressure that is gradually decreased to a point where bubbles start to form, analogous to the situation where production fluids are de-pressured and at some discrete pressure and temperature condition, separate into liquid and gas phases.

Burning velocity: Velocity at which a fuel-air mixture issuing from a burner burns back to the burner.

C1, C2,...,C_n: Shorthand for petroleum hydrocarbon components containing "n" carbon atoms. In its simplest form, C1 denotes methane, C2 denotes ethane, and C3 denotes propane. The larger carbon numbers do not necessarily represent a single component, but rather an aggregation of components that have the same carbon number.

Conditioning: Processing of crude oil to remove impurities prior to transport. Impurities include gases, water and solids that were co-produced with the crude oil.

Deflagration: Classification of an explosion. Burning of a fuel-air mixture where the flame travels at subsonic velocities.

Detonation: Classification of an explosion. Burning of a fuel-air mixture where the flame travels at supersonic velocities.

Equation of State (EOS): Mathematical model relating state variables pressure, density, and temperature. An EOS is useful for interpreting experimental phase behavior data and also predicting P-□-T behavior in process simulations and for conditions where experimental data are not available.

Fireball: Partially pre-mixed diffusion flames which rapidly combust due to enhanced turbulent mixing and atomization.

Flammability limits: Range of vapor concentration in air that will support combustion termed lower flammability limit (LFL) and upper flammability limit (UFL).

Flares: The burning of fuel vapors at the source of a release.

Flash fire: The burning of a fuel vapor cloud which was ignited at a location away from the release point.

Flashpoint: Temperature that results in a vapor concentration in air corresponding to the lower flammability limit. When this temperature is reached there will be a flash of flame without sustained burning. The fire point is the temperature at which sustained burning occurs and is higher than the flashpoint.

Gas-oil ratio (GOR): The volume ratio of gas to liquid evolved from an oil that is depressurized to known P, T conditions. In this study, P = 1 atm, and T = 100°F unless otherwise noted. Volume units are in standard cubic feet of gas per standard barrel of liquid (scf/bbl). (Note that standard conditions for reported gas standard cubic feet per industry standards are P = 1 atm and T= 60°F.)

Initial Boiling Point (IBP): The temperature at which a liquid begins to boil during a distillation process, typically at atmospheric pressure. The result for a liquid containing a wide range of boiling components is highly dependent on method, and there is open debate over which of many available methods is most appropriate for the common application of assigning packing groups for DOT class 3 flammable liquids.

Minimum ignition energy: Minimum energy required to ignite a flammable fuel-air mixture. **Pool fires:** Fires resulting from the burning of liquid fuel pools.

Reid Vapor Pressure (RVP): Measurement of petroleum product vapor pressure per ASTM D323. Measures the equilibrium pressure of an air saturated petroleum product with an air filled vapor chamber with 4 times the volume of the attached liquid filled chamber at a temperature of 100°F. The equilibrium pressure is reported as the Reid Vapor Pressure.

Sour crude oil: Crude oil containing a relatively high mass% sulfur. Standards vary according to context. API (2011) identifies sour oils as having more than 1% total mass sulfur. The U.S. Strategic Petroleum Reserve identifies sour oils as having between 0.5% - 2.0 % total mass sulfur. SPR currently does not accept oils with total mass% sulfur 2.0% or higher.

Stabilization: Removing higher vapor pressure components from a crude oil to lower the volatility of the oil to make it more acceptable for sale and further transport to a refinery or other manufacturing plant.

Sweet crude oil: Crude oil containing a relatively low mass% sulfur. Standards vary according to context. API (2011) identifies sweet oils as having less than 1% total mass sulfur. The U.S. Strategic Petroleum Reserve identifies sweet oils as having less than 0.5% total mass sulfur.

True vapor pressure (TVP): Pressure exerted by a gas that is in thermodynamic equilibrium with a liquid phase. True vapor pressure for pure liquids is typically a known material property and constant at a given temperature. True vapor pressure for mixtures like crude oil is a function of the ratio of gas to liquid volume (V/L) and temperature.

Vapor-liquid equilibrium (VLE): A condition achieved when the chemical potential of each component in a liquid is equalized across the phase boundary, and there is no net transfer of vapor to liquid or liquid to vapor.

***$VPCR_x(T)$**:* Equilibrium vapor pressure over a liquid at vapor/liquid volume ratio = x, and a temperature = T. Measurements should be compliant with ASTM D6377.

In: Crude Oil
Editor: Dianne Gibbs

ISBN: 978-1-63483-070-6

Chapter 2

U.S. RAIL TRANSPORTATION OF CRUDE OIL: BACKGROUND AND ISSUES FOR CONGRESS*

John Frittelli, Anthony Andrews, Paul W. Parfomak, Robert Pirog, Jonathan L. Ramseur and Michael Ratner

SUMMARY

North America is experiencing a boom in crude oil supply, primarily due to growing production in the Canadian oil sands and the recent expansion of shale oil production from the Bakken fields in North Dakota and Montana as well as the Eagle Ford and Permian Basins in Texas. Taken together, these new supplies are fundamentally changing the U.S. oil supply-demand balance. The United States now meets 66% of its crude oil demand from production in North America, displacing imports from overseas and positioning the United States to have excess oil and refined products supplies in some regions.

The rapid expansion of North American oil production has led to significant challenges in transporting crudes efficiently and safely to domestic markets—principally refineries—using the nation's legacy pipeline infrastructure. In the face of continued uncertainty about the prospects for additional pipeline capacity, and as a quicker, more flexible

* This is an edited, reformatted and augmented version of a Congressional Research Service publication R43390, prepared for Members and Committees of Congress, dated December 4, 2014.

alternative to new pipeline projects, North American crude oil producers are increasingly turning to rail as a means of transporting crude supplies to U.S. markets. Railroads are more willing to enter into shorter-term contracts with shippers than pipelines, offering more flexibility in a volatile oil market. According to rail industry officials, U.S. freight railroads delivered 435,560 carloads of crude oil in 2013 (roughly equivalent to 300 million barrels), compared to 9,500 carloads in 2008. In the first half of 2014, 258,541 carloads of crude oil were delivered. Crude imports by rail from Canada have increased more than 20-fold since 2011. The amount of oil transported by rail may also be influenced by a tight market for U.S.-built tankers. However, if recent oil price declines persist and the price falls below the level at which Bakken producers can cover their costs, some production could be shut in, potentially reducing the volume of oil carried by rail.

While oil by rail has demonstrated benefits with respect to the efficient movement of oil from producing regions to market hubs, it has also raised significant concerns about transportation safety and potential impacts to the environment. The most recent data available indicate that railroads consistently spill less crude oil per ton-mile transported than other modes of land transportation. Nonetheless, safety and environmental concerns have been underscored by a series of major accidents across North America involving crude oil transportation by rail—including a catastrophic fire that caused numerous fatalities and destroyed much of Lac Mégantic, Quebec, in 2013. Following that event, the U.S. Department of Transportation issued a safety alert warning that the type of crude oil being transported from the Bakken region may be more flammable than traditional heavy crude oil.

Legislation introduced in Congress following the Lac Mégantic disaster would require railroads to have at least two crew members aboard all trains. In addition, policy makers are proposing regulatory changes involving tank car design, prevention of derailments, and selection of preferred routes for transporting oil by rail. Congress may evaluate these changes in the reauthorization of the Rail Safety Improvement Act of 2008 (P.L. 110-432).

INTRODUCTION

North America is experiencing a boom in crude oil supply, primarily due to the growth of heavy crude production in the Canadian oil sands[1] and the recent expansion of shale oil production in North Dakota, Montana, and Texas. North American production now supplies 66% of U.S. crude oil demand, displacing crude from Latin America, Africa, and the Middle East.

This shift has led to significant challenges in transportation, as refineries that once received crude oil principally from oceangoing tankers are now seeing increasing deliveries by domestic transport. Existing pipeline capacity is, in some cases, insufficient to carry growing crude oil from some production areas, or does not link to the refineries needing the oil. The domestic barge network does not serve some key production regions located far from navigable waterways. As a quicker, more flexible alternative to new pipeline projects, North American crude oil producers are increasingly turning to rail as a means of transporting crude supplies to U.S. markets. Increased exports of refined products—and, if Congress changes the law, of crude oil—could lead to even larger volumes of oil being transported by rail. According to rail industry officials, U.S. freight railroads delivered 435,560 carloads of crude oil in 2013, or roughly 300 million barrels, compared to 9,500 carloads in 2008.[2] In the first half of 2014, 258,541 carloads of crude oil were delivered.[3] Crude imports by rail from Canada have increased more than 20-fold since 2011.

The rapid increase in crude oil shipments by rail will likely increase the number of oil spills from rail transportation. However, the most recent data available indicate that railroads consistently spill less crude oil per ton-mile transported than other modes of land transportation. The amount of crude spilled per ton-mile of rail transport declined significantly between the early 1990s and the 2002-2007 period, the most recent years for which data are available.[4]

Nonetheless, the increase in rail shipments of crude has raised safety and environmental concerns. These concerns have been underscored by a series of major incidents involving crude oil transportation by rail, including a catastrophic fire and explosion in Lac Mégantic, Quebec, in July 2013 and a derailment in Casselton, ND, in December 2013 that led to a mass evacuation. Consequently, government agencies in the United States and Canada have issued new regulations and are considering others related to oil transport by rail, and some Members of Congress have called for tighter rules governing crude oil railcars as well as a broader reconsideration of the role of rail in the nation's oil transportation infrastructure.[5]

Why Is Oil Moving by Rail?

In 2013, the United States produced 2.72 billion barrels of crude oil and imported another 2.82 billion barrels.[6] Canada has become the United States' leading foreign supplier, thanks to its increasing production from oil sands.[7]

However, U.S. oil output has been increasing rapidly. In October 2013, U.S crude oil production exceeded imports for the first time since February 1995.[8]

The location of U.S. crude oil production has been changing rapidly. In particular, production in Alaska and from offshore sites has been declining, while production in Texas and North Dakota has been rising. The U.S. Geological Survey recently estimated that 2.7 billion barrels of light sweet crude oil remain in overlooked producing formations,[9] including the Eagle Ford shale, a prolific source of very light sweet crude oil in Texas, and the Bakken formation in North Dakota, a source of light sweet crude oil that rivals West Texas crude in quality.[10]

Almost all oil produced domestically, as well as some Canadian production, flows to one of the 115 U.S. refineries (*Figure 1).*[11] Nearly 45% of the country's refining capacity is located in the Gulf Coast, where 43 refineries process more than 9 million barrels of oil per day (bpd). However, the Midwest and the West Coast also have significant refining capacity.

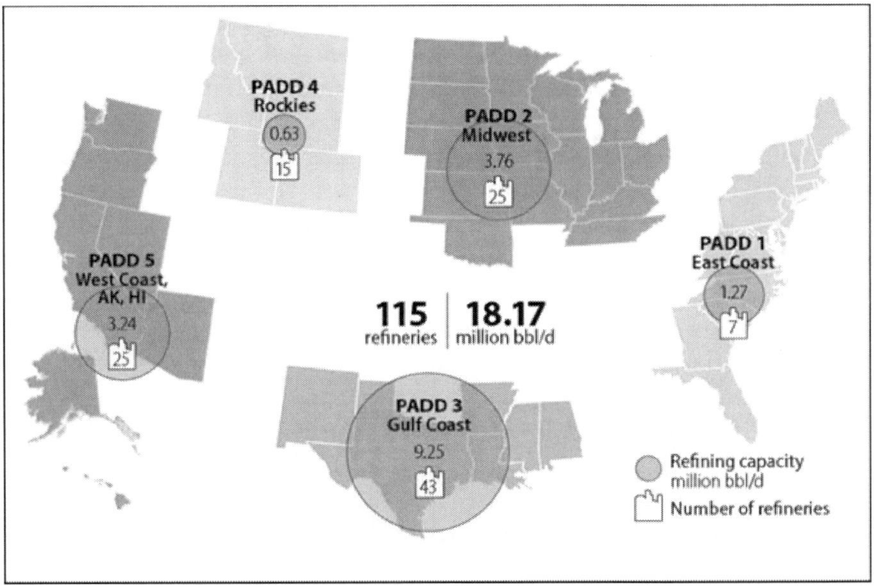

Sources: Congressional Research Service; Energy Information Administration.

Note: PADD = Petroleum Administration for Defense Districts, five districts established by executive order during World War II for gasoline rationing.

Figure 1. U.S. Refinery Capacity by PADD in 2012.

The last entirely new petroleum refinery in the United States opened in 1976. The number of refineries in operation has steadily declined since then as refining capacity has become concentrated in ever larger refineries. A quarter of U.S. capacity is concentrated in 11 refineries with capacities exceeding 300,000 bpd. The largest, Shell/Motiva's Baytown, TX, refinery, was recently expanded to 600,000 bpd. Operable U.S. refining capacity has actually increased from 16.5 million to nearly 18 million bpd over the last decade. Refineries representing approximately 75% of domestic capacity (13.3 million bpd) have the ability to process heavy crude oils, but many smaller refineries can process only light to intermediate crude oil.

Each refinery depends upon a certain grade or blend of crude oils to operate efficiently, depending upon its custom-designed processing equipment. A refinery designed to run light crude oil could not switch to heavy crude oil without adding a coking unit, for example. However some refineries that process heavy sour crude could switch to lighter sweet crude by bypassing their coking units, if the economics of doing so are favorable. Until quite recently, the supply of light sweet crude oil was diminishing, but newly available light sweet crudes from North Dakota's Bakken formation are changing refining dynamics in some regions of the United States, especially as refineries seek supplies that cannot be delivered economically by tanker ships or pipelines.

Traditionally, pipelines and oceangoing tankers have delivered the vast majority of crude to U.S. refineries, accounting for approximately 90% of total receipts (in barrels) in 2013. Although other modes of transportation—rail, barge, and truck—have accounted for a relatively minor portion of crude oil shipments, volumes have been rising very rapidly. The volume of crude oil carried by rail increased by 20-fold between 2008 and 2013, and the volume moving by barge, on inland waterways as well as along intracoastal routes, and by truck doubled during that time frame. Figure 2 shows the change in transportation by mode between 2008 and 2013.

Rail is a relatively high-cost method of transporting oil. Although crude oil transportation costs are typically not a major driver of refiner profitability, refiners are typically wary of incurring any costs that are higher than those faced by their competitors, as all refined petroleum products sold in a region tend to command the same price independent of the refinery that produced them.

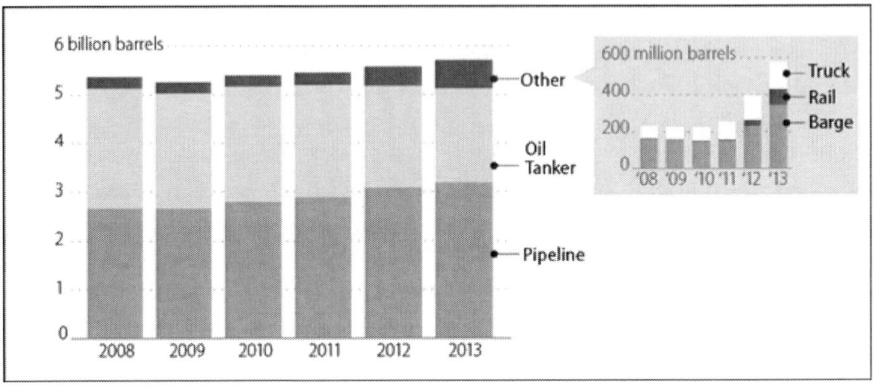

Source: Prepared by CRS; data from EIA, Refinery Capacity Report, Table 9, June 2014.

Notes: Some shipments may involve multiple modes, such as rail to barge. This figure indicates only the mode used for the last leg of such shipments.

Figure 2. U.S. Refinery Receipts of Crude Oil by Mode of Transportation.

The Economics of Oil by Rail

In the short run, rapid expansion of oil production in the Bakken—production volumes increased 10-fold between 2005 and 2014[12]—strained the capacity of existing pipelines and of refiners able to process the oil. Finding ready buyers was difficult, resulting in discounted prices compared to other crude oil traded in the U.S. market. With Bakken crude selling for approximately $4 to $28 per barrel less than West Texas Intermediate (WTI) crude, the U.S. reference price for crude grade, refiners found it profitable to utilize the North Dakota oil delivered by rail even though the rail transportation cost is perhaps $5 to $10 per barrel higher than pipeline costs.

Rail has also been critical to development of Canadian oil sands. Although the vast majority of crude oil imports from Canada are delivered via existing pipeline, imports by rail are estimated to have increased from 1.6 million barrels in 2011 to 40 million barrels in 2013. Construction of the proposed Keystone XL pipeline could move a significant proportion of these shipments off the rails, as pipeline transportation is likely to cost less per barrel.[13]

For certain refiners, the economics of using rail to transport Bakken oil supplies are even more attractive. For example, in 2012, several refineries in the Philadelphia area were scheduled for closure. The refineries were using imported crudes, largely sourced from West Africa, which sold at a premium to WTI.[14] Use of high-cost imported crude oil made their refined products, notably gasoline, uncompetitive against similar products produced by Gulf Coast refineries that used cheaper heavy crudes. By using supplies from the Bakken, these refineries have lowered their costs and have become more competitive. New owners are now investing in the refineries, including installation of high-speed rail unloaders that allow them to use Bakken crude oil. These innovations also reduce the cost of rail transportation per barrel.

The attractiveness of rail transportation of oil may not be permanent. Transporting Bakken crude by rail became cost-effective because of the price discounts created by pipeline bottlenecks, which prevented available supplies from reaching the consuming markets. If additional oil pipeline capacity is constructed to connect North Dakota with consuming markets, the pipeline bottleneck will be reduced.[15] As a greater share of Bakken oil reaches the market via pipeline, it is likely to result in reduced price discounts. With a reduced price discount, Bakken oil would be less competitive in refining when transported by rail. On the other hand, a smaller discount on Bakken crude oil prices would likely lead to greater drilling activity and production from the Bakken fields. Higher production, which expands supply, could support price discounts. Given the uncertainty about the future value of the oil, the longevity of the deposits, environmental issues and the difficulties in permitting and construction, it is uncertain how much pipeline construction will be undertaken. It may be that large volumes of crude oil could be transported by rail well into the future.

Lower oil prices prevalent in the second half of 2014 may also affect Bakken crude. If the price of oil falls to levels lower than the level at which Bakken producers can cover their costs, some production might be shut down, and/or the rate of growth of development and production would likely slow. It is uncertain as to how these changes might affect the price discount on Bakken crude oil.

Railroads are a viable alternative to pipeline transportation largely because they offer greater flexibility. The nation's railroad network is more geographically extensive than the oil pipeline network, and better able to ship crude oil from new areas of production to North American refineries. While there are about 61,000 miles of crude oil pipeline in the United States, there are nearly 140,000 miles of railroad.[16]

The U.S. Railroad Industry in Brief

The U.S. rail network comprises seven large (Class I) railroads, which focus on moving products between North American regions. These railroads generally market to large volume, long-distance shippers. There are also roughly 500 "shortline" (Class II or III) railroads that sometimes serve as the first or final leg of a Class I rail shipment. Shortlines were often spun-off from Class I railroads because of insufficient business over the line. Class I railroads account for about 70% of system mileage, 90% of railroad employees, and about 95% of freight railroad revenue. Since crude oil movements involve non-traditional rail origins (drilling sites) and destinations (refineries), shortlines are often involved in these movements.

Railroad track is categorized into classes that determine the allowable speeds over the track.17 Most track with the lowest speed limits is the property of shortlines. If track needs maintenance work, a railroad will issue a "slow order" on that section of track, reducing train speeds. Class I railroads have transitioned to using bigger and heavier cars, raising the maximum weight on many track sections from 263,000 lbs. to 286,000 lbs. Shortline railroads that interchange traffic with Class I railroads have had to improve their roadbeds to accommodate the heavier cars.

The railroad industry, since 1980, is mostly economically deregulated. The Surface Transportation Board can review the reasonableness of railroad rates and service in situations where the railroad is determined to have "market dominance," generally where a shipper is served by only one railroad and cannot ship economically by other means. As "common carriers," railroads are required to provide rail service upon reasonable request. Railroads do not require a special federal permit to transport crude oil. Federal railroad law preempts state and local authority, which is generally restricted to a state or local government's "police powers."

The geographic flexibility of the railroad network compared to the oil pipeline network can be especially beneficial for a domestic market in flux. Railroads can increase capacity relatively cheaply and quickly by upgrading tracks and roadbeds to accommodate higher train speeds, building passing sidings or parallel tracks, increasing the frequency of switchovers from one track to the other, and upgrading signal systems to reduce the headway needed between trains. Although railroads need approval from the federal Surface Transportation Board (STB) to build new lines, they do not require STB approval to make improvements to existing lines. And even without capacity improvements, railroads can offer routings not served by pipelines.

A significant fall-off in railroad coal movements has increased railroads' capacity to transport oil over some routes. In 2013, railroads carried about 395,000 more tank cars of crude than in 2005, but about 1.3 million fewer cars of coal. To put the increase in crude traffic in perspective, crude oil represented less than 1% of total rail carloads in 2012. In the first half of 2014, crude carloads increased to 1.6% of total rail car loadings. While, on a national scale, increased rail car loadings of crude oil represent a relatively small percentage of total traffic, significant increases in traffic in a specific area can cause bottlenecks that can reverberate across the entire rail network. The STB held hearings in April and September 2014 to hear complaints from non-oil shippers concerning poor rail service in the upper Midwest due to oil traffic and the severe winter weather.[18] The STB ordered BNSF and CP railroads to report how they intended to ensure delivery of fertilizer to farmers in spring 2014, and required all the railroads to provide weekly status reports on their performance. At the hearing, BNSF (the railroad most directly serving the Bakken region) noted that its car loadings in North Dakota had more than doubled from 2009 to 2013, and that in October 2013, crude oil and agricultural car loadings surged by more than it could manage. Past experience has shown that railroad bottlenecks are not quickly resolved.

This experience illustrates that pipelines can generally provide more reliable service than railroads. Among other differences, rail shipments are more affected by weather. In addition, railroads generally experience peak demand during the fall due to the grain harvest and retailers' holiday shipments. This may cause locomotives and track capacity to be in shorter supply at certain times of the year.

Railroad transport reportedly costs in the neighborhood of $10 to $15 per barrel compared with $5 per barrel for pipeline. In return, railroads offer oil producers certain advantages. Heated railroad tank cars improve the viscosity of oil sands crude so that less diluent needs to be added than if the product were being moved by pipeline. Generally, railroads are more willing to enter into shorter-term contracts with shippers than pipelines (1 to 2 years versus 10 to 15 years), offering more flexibility in a rapidly changing oil market. Recent oil price volatility seems to underscore this advantage. Moving oil by train from North Dakota to the Gulf Coast or Atlantic Coast requires about 5 to 7 days' transit, versus about 40 days for oil moving by pipeline, reducing producers' need for working capital to cover the cost of oil in transit.[19]

Crude oil often moves by unit train, a train that carries just one type of cargo in a single type of car and serving a single destination. Unit trains do not need to be switched or shunted in rail yards, saving time and reducing costs, and return to their point of origin as soon as they have been unloaded. A train consisting of 70 to 120 tank cars can carry in the neighborhood of 50,000 to 90,000 barrels of oil, depending on the type of crude.

One hindrance to the expansion of crude-by-rail has been the lack of tank cars and loading and unloading infrastructure. Much of this investment is being made by the oil industry or by rail equipment leasing companies, not railroads. As of April 2014, manufacturers had 50,000 crude oil tank cars on order, on top of an existing fleet of 43,000. (This is in addition to 30,000 tank cars that carry ethanol and 27,000 that carry other flammable liquids.) In 2013, over 28,000 tank cars of all types were built, up from over 17,000 in 2012.[20] Facilities for building tank cars are unique because the process involves baking the entire car in an oven. One manufacturer believes the tank car builders are capable of increasing production each year by 7,000 to 10,000 cars.

Rail terminal capacity is expected to increase fourfold from 2012 to 2015.[21] Matching the daily throughput volume of a pipeline requires several trains per day, with each train taking 13 to 24 hours to unload; oil rail terminals therefore require large areas for parallel loop tracks where multiple trains can await unloading.

The Role of Barges and Ships in Domestic Crude Transportation

Many refineries traditionally have received crude from overseas and thus are located near the coastline with access to dock facilities. Some are not equipped to receive crude by rail. Hence, some railroads are transferring oil to barges for the last leg of the trip to refineries. Locations where railroads transfer crude oil to barges include St. Louis and Hayti, MO; Osceola, AR; Hennepin, IL; Albany, NY; Yorktown, VA; and Anacortes and Vancouver, WA. In addition, crude produced at Eagle Ford, TX, which is located near ports, is being moved along the coast by either barge or ship.

One river barge can hold 10,000 to 30,000 barrels of oil. Two to three river barges are typically tied together in a single tow that carries 20,000 to 90,000 barrels, about the same load as a unit train. Coastal tank barges designed for open seas, known as articulated tug-barges, or ATBs,[22] can hold 50,000 to 185,000 barrels, although newer ATBs can carry as much as 340,000 barrels, comparable to the capacity of coastal tankers. ATBs are slower, less fuel-efficient, and more restricted by sea conditions, but nevertheless may have an economic advantage over tankers because Coast Guard crewing regulations allow them to sail with one-third to half the crew required on a tanker. Crude oil tankers used to move Alaska oil to West Coast refineries have capacities of 800,000 to over 1 million barrels.

An advantage of tankers over railroads is the greater amount of oil they can carry in a single voyage, which better matches the daily consumption rate of refineries. With the median capacity for U.S. refineries at about 160,000 barrels per day, a coastal tanker can carry a two-day supply of oil. In addition, while railroads must build and maintain tracks and pay property taxes on their rights-of-way, the ocean is free, and harbor channels are largely provided by the federal government. For these reasons, tankers can be much cheaper than railroads in moving oil, even though the railroad route may be much more direct. For instance, the distance between the Bakken region in North Dakota and refineries in the Northeast is approximately 1,800 miles, and the cost of railroad transport is $14 per barrel.[23] The distance from Texas ports near the Eagle Ford region to the same refineries is about 2,100 miles, and tanker rates are $5 to $6 per barrel.[24] Similarly, the overland distance from the Eagle Ford region to Los Angeles-area refineries is about 1,400 miles, and the estimated cost of railroad transport is $15 per barrel, while the water route through the Panama Canal is 5,200 miles and is estimated to cost $10 per barrel.[25]

Although seemingly a circuitous route compared to rail, it is not inconceivable that tankers could play a role in moving Bakken oil to East or West Coast refineries. Significant amounts of Bakken oil are moved to Gulf Coast terminals by pipeline, railroad, barge, or combinations of these modes for refining within that region. From a Gulf Coast port, tankers could transport the oil to either East or West Coast refineries. Via existing rail and pipeline connections to Great Lakes ports, tankers could also move Bakken oil from there to Northeast refineries. However, the economic viability of these routes, in particular, and routes involving domestic coastal transport in general, is heavily influenced by the Jones Act.[26]

The Jones Act

The Jones Act may have a profound impact on where crude oil is sourced and how it is transported. The Jones Act requires that vessels transporting cargo between two U.S. points be built in the United States, as well as crewed and at least 75% owned by U.S. citizens.[27] The domestic build requirement for tanker ships, in particular, has been identified as contributing to higher costs in moving domestic crude oil along the coasts.[28] Domestically built tankers are about four times the price of foreign-built tankers,[29] and there is limited capacity in U.S. shipyards to build them. Much of the existing crude oil tanker fleet was built since 2000 to meet Oil Pollution Act of 1990 (P.L. 101-380) requirements that tankers calling at U.S. ports have double hulls. Two crude carriers were delivered in 2014 to replace two vessels in Alaska trade.

The Jones Act-eligible fleet of crude oil tankers consists of 10 ships, all employed in moving Alaska crude oil to the U.S. West Coast or to a refinery in Alaska.[30] Since annual Alaska oil production has fallen by about 46% over the last decade, the Jones Act crude oil fleet has been in decline. About 30 Jones Act-eligible tankers carry chemicals or refined petroleum products, such as gasoline or jet fuel, but these ships do not readily alternate between carrying dirty oil (crude oil, residual fuel oil, asphalt) and refined (clean) petroleum products because the tanks would have to be extensively washed after carrying dirty product, a time-consuming and costly process. Some product vessels have fundamentally different designs from crude carriers and would require a layup in a shipyard to be converted to move crude oil.[31]

Phillips 66 has chartered two Jones Act tankers to move crude oil from Eagle Ford, TX, to a refinery in Linden, NJ (in proximity to New York Harbor).[32] Phillips 66 has stated that if more Jones Act-eligible tankers were available, it would like to receive 100,000 barrels a day of Eagle Ford oil at this refinery (it would need several tankers to accomplish this, the exact number depending on the size of the tankers).[33] EIA data (which specify oil movement only between regions, not to individual refineries) indicate that roughly five times more Texas crude oil is being shipped in foreign-flag tankers to refineries in eastern Canada than is being shipped in Jones Act-qualified tankers to U.S. Northeast refineries.[34] The oil shipments from Texas to Canada cost approximately $2 per barrel, compared with $5 to $6 per barrel for shipments from Texas to U.S. Northeast refineries. The cost difference for a 300,000-barrel tanker amounts to around $1 million, meaning that a Texas oil producer receives $1 million less for its oil when shipping to U.S. Northeast refineries than when shipping to Canadian refineries.[35] The Bayway refinery also receives 50,000 barrels of Bakken oil per day by rail, and is finishing construction of a rail unloading terminal with capacity of 75,000 barrels per day. The refinery has a capacity of 238,000 barrels per day. It may also source offshore oil from eastern Canada (with shipping rates fluctuating around $1.20 per barrel), as EIA data indicate that this was a top source of foreign oil for Northeast refineries in 2013.[36]

The Role of Tank Trucks

Tank trucks operating on U.S. roadways have been an important link in moving crude oil from domestic drilling sites to pipelines and rail terminals. A typical tank truck can hold 200 to 250 barrels of crude oil. Trucks readily serve the need for gathering product, as the hydraulic fracturing method of drilling employed in tight oil production involves multiple drilling sites in an area and the location of active wells is constantly in flux. A large volume of crude oil is being transported by truck between production areas and refineries in Texas because of the close proximity of the two.

OIL SPILL CONCERNS

Each mode of oil transportation—pipelines, vessels, rail, and tanker trucks—involves some risk of oil spills. Over the period 1996-2007, railroads consistently spilled less crude oil per ton-mile than trucks or pipelines. Barges and domestic tanker ships have much lower spillage rates than trains (*Figure 3*). However, the data in *Figure 3* precede the recent dramatic increase in oil transportation by rail.

Given the comparatively small capacity of a rail tank car, around 700 barrels, the total amount spilled from even a major derailment is likely to be small compared to the 260,000 barrels discharged in the 1989 grounding of the *Exxon Valdez* in Prince William Sound, AK, or the approximately 40,000 barrels discharged in the largest U.S. pipeline oil spill CRS can document, which occurred in 1991 near Grand Rapids, MN.[37] Nonetheless, spill volume is arguably a relatively unimportant factor in terms of impacts and cleanup costs. Location matters more: a major spill away from shore will likely cost considerably less to abate than a minor spill in a populated location or sensitive ecosystem. Depending on timing and location, even a small spill can cause significant harm to individual organisms and entire populations.[38]

Although spillage per ton-mile of oil transported by rail declined over time, a recent series of major accidents (see the text box) has heightened concern about the risks involved in shipping crude by rail.

The increasing deployment of unit trains changes the risks involved in shipping oil by rail in two ways. Unit trains of crude oil concentrate a large amount of potentially environmentally harmful and flammable material, increasing the probability that, should an accident occur, large fires and explosions could result.

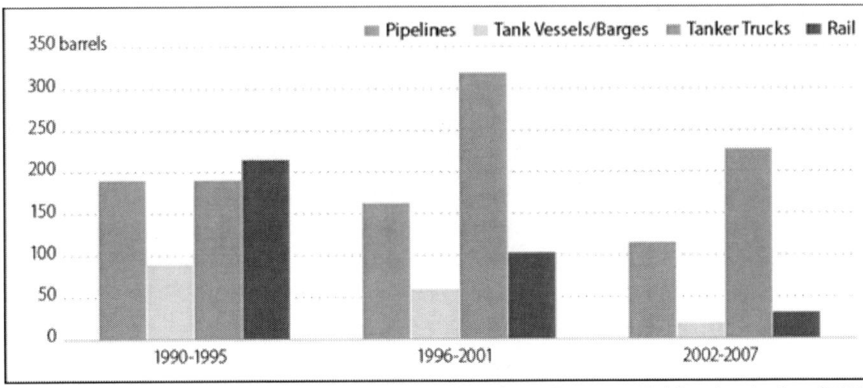

Sources: Prepared by CRS; oil spill volume data from Dagmar Etkin, Analysis of U.S. Oil Spillage, API Publication 356, August 2009; ton-mile data from Association of Oil Pipelines, Report on Shifts in Petroleum Transportation: 1990-2009, February 2012.

Notes: Pipelines include onshore and offshore pipelines. The time periods were chosen based on the available annual data for both spill volume and ton-miles. The values for each time period are averages of annual data for each six-year period.

Figure 3. Crude Oil and Petroleum Product Spills during Domestic Transportation. (Oil Spill Volume (Barrels) per Billion-Ton-Miles)

This risk is similar to that of unit trains carrying ethanol, and maybe greater than that of mixed freight trains in which various hazardous materials, such as explosives and toxic-by-inhalation materials, are sequenced among other cars according to federal regulations.39 On the other hand, while unit trains concentrate a voluminous quantity of potentially dangerous material, they may offer safety benefits from avoiding the decoupling and re-coupling of cars in rail yards, which involve high-impact forces and introduce opportunity for human error.

Oil by Rail Derailments in 2013 and 2014

Lac Mégantic, Quebec—On July 5, 2013, a train with 72 loaded tank cars of crude oil from North Dakota moving from Montreal, Quebec, to St. John, New Brunswick, stopped at Nantes, Quebec, at 11:00 pm. The operator and sole railroad employee aboard the train secured it and departed, leaving the train on shortline track with a descending grade of about 1.2%. At about 1:00 AM, it appears the train began rolling down the descending grade toward the town of Lac-Mégantic, about 30 miles from the U.S. border. Near the center of town, 63 tank cars derailed, resulting in multiple explosions and subsequent fires. There were 47 fatalities and extensive damage to the town. 2,000 people were evacuated. The initial determination was that the braking force applied to the train was insufficient to hold it on the 1.2% grade and that the crude oil released was more volatile than expected.

Gainford, Alberta—On October 19, 2013, nine tank cars of propane and four tank cars of crude oil from Canada derailed as a Canadian National train was entering a siding at 22 miles per hour. About 100 residents were evacuated. Three of the propane cars burned, but the tank cars carrying oil were pushed away and did not burn. No one was injured or killed. The cause of the derailment is under investigation.

Aliceville, Alabama—On November 8, 2013, a train hauling 90 cars of crude oil from North Dakota to a refinery near Mobile, AL, derailed on a section of track through a wetland near Aliceville, AL. Thirty tank cars derailed and some dozen of these burned. No one was injured or killed. The derailment occurred on a shortline railroad's track that had been inspected a few days earlier. The train was travelling under the speed limit for this track. The cause of the derailment is under investigation.

Casselton, North Dakota—On December 30, 2013, an eastbound BNSF Railway train hauling 106 tank cars of crude oil struck a westbound train carrying grain that shortly before had derailed onto the eastbound track. Some 34 cars from both trains derailed, including 20 cars carrying crude, which exploded and burned for over 24 hours. About 1,400 residents of Casselton were evacuated but no injuries were reported. The cause of the derailments and subsequent fire is under investigation.

Plaster Rock, New Brunswick—On January 7, 2014, 17 cars of a mixed train hauling crude oil, propane, and other goods derailed likely due to a sudden wheel or axle failure. Five tank cars carrying crude oil caught fire and exploded. The train reportedly was delivering crude from Manitoba and Alberta to the Irving Oil refinery in Saint John, New Brunswick. About 45 homes were evacuated but no injuries were reported.

Philadelphia, Pennsylvania—On January 20, 2014, 7 cars of a 101-car CSX train, including 6 carrying crude oil, derailed on a bridge over the Schuylkill River. No injuries and no leakage were reported, but press photographs showed two cars, one a tanker, leaning over the river.

Vandergrift, Pennsylvania—On February 13, 2014, 21 tank cars of a 120-car train derailed outside Pittsburgh. Nineteen of the derailed cars were carrying crude oil from western Canada, and four of them released product.

There was no fire or injuries.

Lynchburg, Virginia—On April 30, 2014, 15 cars in a crude oil train derailed in the downtown area of this city. Three cars caught fire, and some cars derailed into a river along the tracks. The immediate area surrounding the derailment was evacuated. No injuries were reported.

In March and April 2013, there were two derailments of Canadian Pacific trains, one in western Minnesota and the other in Ontario, Canada; less than a tank car of oil leaked in each derailment and neither incident caused a fire.

The increasing deployment of unit trains changes the risks involved in shipping oil by rail in two ways. Unit trains of crude oil concentrate a large amount of potentially environmentally harmful and flammable material, increasing the probability that, should an accident occur, large fires and explosions could result. This risk is similar to that of unit trains carrying ethanol, and maybe greater than that of mixed freight trains in which various hazardous materials, such as explosives and toxic-by-inhalation materials, are sequenced among other cars according to federal regulations.[39] On the other hand, while unit trains concentrate a voluminous quantity of potentially dangerous material, they may offer safety benefits from avoiding the decoupling and re-coupling of cars in rail yards, which involve high-impact forces and introduce opportunity for human error.

Special Concerns About Canadian Dilbit

Oil companies generate substantial quantities of crude oil and related substances from the natural bitumen in oil sands, particularly deposits in Alberta, Canada. In 2012, the United States imported 438 million barrels of oil sands-derived crude oils, 125% more than in 2005.[40] Because bitumen is highly viscous, it is transported mostly in the form of diluted bitumen, or dilbit, containing naptha or other materials that make it flow more easily.

Some commenters have argued that due to its physical characteristics, dilbit presents greater risks of oil spills than conventional crude, with potentially greater impacts to the environment.[41] Other stakeholders and organizations have questioned these conclusions.[42] A study released by the National Research Council in 2013, conducted at the direction of Congress,[43] found that the characteristics of dilbit do not increase the likelihood of spills.[44] The extent to which these findings are applicable to rail transport of crude is open to debate, as rail tanker cars may have different operating parameters (e.g., temperature) and physical standards (e.g., wall thickness), or may transport different forms of oil sands-derived crude oil, decreasing the relevance of the NRC findings.

However, observations in the aftermath of a 2010 pipeline spill are consistent with the assertion that dilbit may pose different hazards, and possibly different risks, than other forms of crude oil. On July 26, 2010, a pipeline owned by Enbridge Inc. released approximately 850,000 gallons of dilbit into Talmadge Creek, a waterway that flows into the Kalamazoo River in Michigan.[45] Three years after the spill, response activities continued,[46] because, according to EPA, the oil sands crude "will not appreciably biodegrade."[47] The dilbit sank to the river bottom, where it mixed with sediment, and EPA has ordered Enbridge to dredge the river to remove the oiled sediment.[48] As a result of this order, Enbridge estimated in September 2013 its response costs would be approximately $1.2 billion,[49] which is substantially higher than the average cost of cleaning up a similar amount of conventional oil.[50]

Special Concerns About Bakken Crude

The properties of Bakken shale oil are highly variable, even within the same oil field. In general, however, Bakken crude oil is much more volatile than other types of crude.[51] Its higher volatility may have important safety implications.[52]

In January 2014, the Pipeline and Hazardous Materials Safety Administration (PHMSA) within the U.S. Department of Transportation (DOT) issued a safety alert warning that recent derailments and resulting fires indicate that crude oil being transported from the Bakken region may be more flammable than traditional heavy crude oil.[53] PHMSA, whose rules are enforced by the Federal Railroad Administration with respect to railroads, reinforced the requirement to properly test, characterize, classify, and where appropriate sufficiently degasify hazardous materials prior to and during transportation. Under its initiative "Operation Classification," PHMSA is to continue to collect samples and measure the characteristics of Bakken crude as well as oil from other locations. In November 2014, the North Dakota Department of Mineral Resources proposed oil conditioning and processing standards before the oil is shipped to improve transportation safety.[54]

Federal Oversight of Oil Transport by Rail

The Federal Railroad Administration (FRA) has jurisdiction over railroad safety. It has about 500 federal inspectors throughout the country and also utilizes about 180 state railroad safety inspectors.55 State inspectors predominantly enforce federal requirements because federal rail safety law preempts state law, and federal law is pervasive. The FRA uses past incident data to determine where its inspection activity should be targeted, although the FRA Administrator stated that in light of the growth of crude-by-rail transportation, the agency also must look for "pockets of risk."56 FRA regulations cover the safety of track, grade crossings, rail equipment, operating practices, and movement of hazardous materials (hazmat). The Pipeline and Hazardous Materials Safety Administration within DOT (PHMSA) issues requirements for the safe transport of hazmat by all modes of transportation, which the FRA enforces with respect to railroads.[57]

Rail incidents are investigated by the National Transportation Safety Board (NTSB), an independent federal agency. The NTSB makes recommendations toward preventing future incidents based on its findings.[58] Unlike the FRA, the NTSB is not required to weigh the costs against the benefits when considering additional safety measures and it has no regulatory authority. Many of the NTSB's recommendations concerning oil transport by rail are identical to those it previously issued for transporting ethanol by rail. While the FRA has largely agreed with NTSB's recommendations, its rulemaking process involves consultation with industry advisory committees, and it must determine which of the many rail safety measures under evaluation deserve priority. Implementing a change in FRA regulations can take years.

U.S. safety requirements apply to any train operating in the United States, regardless of its origin or destination. Canadian safety regulations are very similar but do not exactly mirror U.S. requirements. Cross-border shipments must meet the requirements of both countries. Safety standards established by the rail industry, which often exceed government requirements, apply to both U.S. and Canadian railroads.

When a rail incident results in the release of oil, state, territorial, or local officials are typically the first government representatives to arrive at the scene and initiate immediate safety measures to protect the public. The National Oil and Hazardous Substances Pollution Contingency Plan, often referred to as the National Contingency Plan (NCP), indicates that state, territorial, or local officials may be responsible for conducting evacuations of affected populations. These first responders also may notify the National Response

Center to elevate an incident for federal involvement, at which point the coordinating framework of the NCP would be applied.

Unlike most federal emergency response plans, which are administrative mechanisms, the NCP is codified in federal regulation and is binding and enforceable.[59] The NCP regulations apply to applicable spills from vessels, pipelines, onshore facilities, and offshore facilities. The definition of "onshore facility" includes, but is not limited to "motor vehicles and rolling stock."[60]

If an oil discharge affects navigable waterways, shorelines, or "natural resources belonging to, appertaining to, or under the exclusive management authority of the United States,"[61] Section 311 of the Clean Water Act, as amended by the Oil Pollution Act of 1990, Section 311(c), provides explicit federal authority to respond.[62] The term "discharge" is defined broadly and is not linked to specific sources of oil. The President has the authority to perform cleanup immediately using federal resources, monitor the response efforts of the spiller, or direct the spiller's cleanup activities.[63] Several executive orders have delegated the President's response authority to the Environmental Protection Agency (EPA) within the "inland zone" and to the U.S. Coast Guard within the coastal zone, unless the two agencies agree otherwise.[64] The lead federal agency serves as the On-Scene Coordinator to direct the federal resources used in a federal response.

Regulations require that railroads have either a so-called "basic" response plan or a more "comprehensive" response plan, depending on the volume capacity of the rail car transporting the oil.[65] Comprehensive plans are subject to FRA approval, and must ensure by contract or other means that personnel and equipment are able to handle a worst-case discharge.[66] However, the regulatory threshold for the comprehensive response plan is a tank car holding more than 1,000 barrels, so does not apply to the DOT-111 tank cars used today, which hold around 700 barrels of oil apiece. For these smaller tank cars, railroads must prepare only "basic" response plans, which are not subject to FRA approval.

This threshold was established in 1996,[67] before the advent of oil unit trains, each of which may transport, in aggregate, approximately 70,000 barrels (almost 3 million gallons) of oil. The NTSB has recommended that the threshold for comprehensive plans be lowered to take into account the use of unit trains.[68] On August 1, 2014, U.S. DOT issued an advance notice of proposed rulemaking lowering the threshold.[69]

ISSUES FOR CONGRESS

While oil by rail has demonstrated benefits with respect to the efficient movement of oil from producing regions to market hubs, the dramatic increase in oil by rail shipments has generated interest in several related issues. These include railroad safety, environmental concerns, and tradeoffs over rail versus pipeline development.

Railroad Safety and Incident Response

Prior to the Lac Mégantic derailment, the FRA had increased its inspection activity with regard to trains carrying crude oil. After the derailment, the FRA and PHMSA (along with Transport Canada) initiated a comprehensive review of safety requirements.[70] Three areas of active discussion involve tank car design, prevention of derailments, and railroad operations. Railroads are an integrated system of fixed infrastructure, rolling equipment, and workers. Railroad safety experts note that improving safety performance requires recognition of the parameters and interactive effects among these components, and thus approaching railroad safety as an optimization problem is appropriate.[71] For instance, each additional safety feature on tank cars may increase their weight. This reduces the amount of product carried in each car due to track and bridge weight limits, potentially requiring additional tank cars to carry the same amount of product, and thus increasing the expected number of accidents.

In February 2014, the U.S. DOT reached an agreement with railroads under which they would voluntarily take measures to improve the safety of oil trains.[72] These measures include adding braking power; reducing train speeds to 40 mph through certain urban areas starting July 1 for trains with at least 20 tank cars of crude oil and at least one tank car of the older DOT-111 standard; installing additional wayside wheel bearing detectors by July 1; and other actions. Some of the measures the railroads agreed to are similar to those that the industry already takes for trains carrying "toxic-by-inhalation" hazardous materials. They also are similar to those the U.S. DOT proposed in its August 2014 notice of proposed rulemaking (NPRM) to increase the safety of oil trains.[73]

Tank Car Safety Design

The U.S. DOT establishes construction standards for tank cars.[74] A tank car used for oil transport is roughly 60 feet long, about 11 feet wide, and 16 feet high (see *Figure 4*). It weighs 80,000 pounds empty and 286,000 pounds when full. It can hold about 30,000 gallons or 715 barrels of oil, depending on the oil's density. The tank is made of steel plate, 7/16 of an inch thick (see 49 C.F.R. §179.201).[75] An oil tank car is typically loaded from the top valve and unloaded from the bottom valve. Loading or unloading each car may take several hours, but multiple cars in a train can be loaded or unloaded simultaneously.

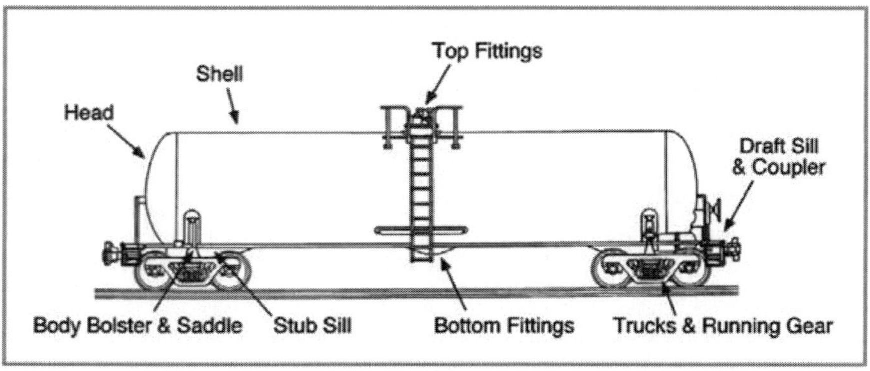

Figure 4. Non-jacketed, Non-pressure Tank Car.

In some incidents, oil has been released from tank cars because the coupler from a neighboring car punctured the tank during derailment. Valves at the top and bottom of the cars have also been sheared off or otherwise opened during derailment. Efforts to improve crashworthiness have focused on reinforcing the shells of tank cars at both ends and/or along the sides with a "jacket" of steel, adding protective head shields at both ends, modifying couplers to prevent decoupling, adding skid protection or diversion shields to protruding valves and modifying pressure relief valves, eliminating or modifying bottom valves so that handles break off rather than opening the valve during derailment, and increasing insulation for fire protection.[76]

The FRA and PHMSA have questioned whether Bakken crude oil, given its characteristics, would more properly be carried in tank cars that have additional safety features, such as those found on pressurized tank cars used for hauling explosive liquids.[77] The railroad industry established additional standards in October 2011 for newly built cars that address some but not all of the safety features that the FRA and PHMSA are considering. Railroads, shippers, and tank car builders have not reached agreement on a replacement standard. Rail cars have an economic life of 30 to 40 years, so conversion of the fleet to a new car standard could take some time.

DOT has asked for further information on the costs and benefits of retrofitting the existing fleet.[78] In November 2013, the Association of American Railroads stated it supports either retrofitting or phasing out oil tank cars built before October 2011 (a fleet of about 78,000 cars) and modifying those built after October 2011 (about 14,000 cars).[79] Some Members of Congress have urged DOT to expedite the rulemaking process concerning new tank car safety requirements.[80]

In February 2014, BNSF announced that it would purchase 5,000 tank cars with safety features that surpass those specified in the October 2011 industry standard (shell thickness would be 9/16 of an inch thick, and they would be jacketed). In April 2014, Transport Canada announced that the oldest DOT-111 tank cars (about 5,000 that lack bottom reinforcement) would no longer be allowed for use in transporting dangerous goods, and the remaining fleet would either be phased out or retrofitted within three years.[81]

In its August 1, 2014, NPRM,[82] the U.S. DOT proposed three options for tank cars built after October 1, 2015: (1) construct with 9/16 inch steel and equip with electronically controlled pneumatic (ECP) brakes (an advanced braking technology), (2) construct with 9/16 inch steel but not with ECP brakes, or (3) construct to the October 2011 industry standard and adopt this as the regulatory standard also. Comments filed in response to the NPRM indicate, among other things, that there is substantial disagreement over the rate at which older tank cars could be retrofitted.[83] U.S. railroads contend that U.S. DOT's proposal is not in harmony with Canada's proposed tank car standards.[84]

Preventing Derailments

An analysis of freight train derailments from 2001 to 2010 on Class I railroads' mainline track found that broken rails or track welds were the leading cause of derailments, by far.[85] These problems caused 670 derailments over the period, while the next leading problem (track geometry defects) caused just over 300 derailments. Broken rails or welds also resulted in more severe incidents, derailing an average of 13 railroad cars instead of 8.6 cars for all other causes. Broken rails or welds accounted for 23% of total cars derailed. A separate study covering the same time period found that track problems were the most important causes of derailments, followed by problems with train equipment.[86]

In the Rail Safety Improvement Act of 2008 (P.L. 110-432, Section 403(a)), Congress requested that the FRA study and consider revising the frequency and methods of track inspection. The FRA conducted the study and on January 24, 2014 issued a final rule on improving rail integrity.[87] The new rule requires railroads to achieve a specified track failure rate rather than scheduling inspections based on the calendar or traffic volume. It also allows railroads to maximize use of rail inspection vehicle time by prioritizing remedial action when track defects are detected. While the rule requires railroads to achieve a lower track failure rate for track that has higher speed limits, or carries passengers or hazardous material, it does not require lower failure rates for track travelling through populated areas or otherwise consider the affected area around the track in the event of a derailment.[88]

The final rule states that it “codifies standard industry good practices,” and notes that railroads “were already initiating and implementing the development of a performance-based risk management concept for determination of rail inspection frequency,” meaning that railroads generally have been testing more frequently than required.[89] Large railroads divide their network into hundreds of segments, and determine inspection frequency for each segment based on past inspection results, past history of undetected defects, track tonnage, climate (namely temperature), signaling system (or absence) over segment, whether track carries toxic-byinhalation materials, and whether track is within 500 feet of a structure.[90] They have vehicles to detect defects within the steel of the rail, and others that detect defects in track geometry. From 1980 to 2012, railroads reduced the number of accidents releasing hazmat product per 100,000 hazmat carloads from 14 to 1.[91]

As part of the February 2014 agreement with DOT referenced above, railroads will perform one additional internal-rail inspection each year than required by the FRA on routes over which trains carrying 20 or more tank cars of crude oil travel, and will conduct at least two high-tech track geometry inspections over these routes. Congress may want to look into research on track defect detection technology, including the feasibility of installing equipment on locomotives to achieve near-continuous rail testing.[92]

Shortline Track

It is often the case that a Class I railroad, prior to turning over the operation of a line to a shortline, did not maintain it to the same standards as busier mainlines. Shippers using a shortline often do not require higher-speed track because they ship infrequently or because the commodities they ship are not time-sensitive. Thus, shortline track is frequently maintained at a lower standard than Class I railroads' track. The Lac Mégantic, Quebec, and Aliceville, AL, crude oil derailments occurred on shortline track. Members of Congress have been concerned with preserving shortline rail service, reflected in a federal loan program for track rehabilitation and improvement and a tax credit for shortline track maintenance.[93]

Railroad Operations

A number of specific operational issues have been found relevant to railroad safety, in general, or to oil by rail transportation specifically.

Train Speed and Braking Systems

Reducing train speed can reduce the number of cars that derail, as well as the likelihood that product will be released.[94] In its August 1, 2014 NPRM,[95] the U.S. DOT proposed that crude oil unit trains be restricted to 50 mph if all tank cars in the train comply with the proposed new standard. For cars not complying with the new standard, speed would be restricted to 40 mph either in all areas or just in urban areas. Trains without enhanced braking systems would be restricted to 30 mph, an accommodation for shortline railroads. Railroads oppose the proposed speed restrictions, arguing that they would severely degrade service for other trains.[96]

In its August 1 NPRM, the U.S. DOT also proposed that crude oil unit trains either use cars equipped with ECP brakes, as mentioned above, or contain another locomotive toward the rear of the train (remotely controlled by the crew in the lead locomotive) or an end of train device (EOT), which are two other methods for sending brake signal commands faster to all the cars in a train. The intent is to reduce the speed of cars upon derailment, but the latter two methods are more effective when the derailment occurs toward the front of the train.[97] Railroads contend that the costs of ECP brakes far outweigh the benefits.[98]

Railroad Crew Size

Following the Lac Mégantic disaster, legislation (H.R. 3040) was introduced in Congress to require two-person crews on all trains. In the United States, the FRA does not specify in regulation how many persons must operate a train, but notes that the various tasks required while a train is moving essentially necessitate at least a two-person crew. Most trains operate with an engineer and a conductor, but some shortline railroads may operate trains with a single crew member. In April 2014, the FRA announced it intended to issue a proposed rule requiring two-person crews (allowing for some exceptions), but this has not occurred as of yet.[99] One potential trade-off is that distraction by a fellow crew member has been found to be a factor in past accidents.[100]

Positive Train Control

Railroads are in the process of implementing positive train control (PTC), a system that is designed to override human error in controlling the speed and movement of trains. Congress required that this system be installed on routes carrying passengers or poison- or toxic-by-inhalation hazardous materials (Section 104 of P.L. 110-432), a requirement that applies to about 60,000 miles of railroad. Current law does not require installation of PTC solely because a track carries crude oil, but the law authorizes the FRA to expand the scope of tracks required to have PTC. PTC is not required on track in or near rail yards. The cost and timeline for implementing PTC are topics of current debate among policy makers and stakeholders.[101]

Route Selection

In the Implementing Recommendations of the 9/11 Commission Act of 2007 (P.L. 110-53, Section 1551), Congress required railroads carrying certain kinds and quantities of potentially dangerous commodities to assess the safest and most secure routes for trains carrying these products and to minimize delays and storage for rail cars containing these products. These requirements currently apply to explosive, toxic-by-inhalation, and radioactive material.[102] Security regulations also require that rail cars containing these commodities not be left unattended when being transferred from one carrier to another or between carrier and shipper.[103] The law resulted from efforts by cities like Washington, DC, and Pittsburgh to ban trains carrying hazardous materials.[104] The FRA is proposing that this routing analysis also apply to unit trains of crude oil.[105] As part of the February 2014 agreement with DOT mentioned above, railroads agreed to perform this routing analysis for oil trains beginning July 1, 2014.

Incident and Oil Spill Response

The increased use of rail for crude oil shipments is likely to increase the number of incidents, some of which may involve oil spills. As described above, the National Oil and Hazardous Substances Contingency Plan provides a framework for federal, state and local collaboration in response to releases of oil and hazardous substances. Considering the relative proximity of rail shipments to population centers, a potential issue for Congress is the safety and adequacy of spill response.

In addition, based on past history, increased frequency or severity of incidents related to shipments of crude oil by rail could lead some local communities to seek additional funding to ensure adequate spill response capabilities, including personnel, training, equipment, and community notification.

The Accuracy of Train Cargo Information

Crude oil may sometimes be carried by "mixed trains"—trains carrying a variety of different commodities. With mixed trains, it is important to first responders that they have an accurate list of which cars contain what commodities (the train "consist"). Often the sequencing of cars changes en route, so the consist information provided by the crew at the scene of an incident may no longer be accurate. Although all vehicles containing hazardous materials must display placards indicating their potential dangerous characteristics (e.g., flammable, corrosive, explosive), responders often need more specific information about the commodities involved in an incident. One potential remedy under consideration is an electronic manifest system that would offer the capability of easier updates. In MAP-21, Congress authorized PHMSA to conduct pilot projects on paperless hazmat information sharing among carriers (of various modes including rail) and first responders.[106] A potential drawback raised by the railroads is that electronic devices at the scene of an incident could encounter technical problems. Another remedy is greater diligence by railroad crew in keeping the paper consist up to date. The NTSB has asked whether a copy of the consist should also be kept at the end of a train in case the copy kept by the crew at the head of the train is lost in an incident.

Rail vs. Pipeline Development

Certain rail routings of crude oil could be replaced by reconfiguring the existing pipeline network and constructing additional pipeline capacity. In general, pipelines could provide safer, less expensive transportation than railroads, assuming that pipeline developers are able to assure markets for the oil they hope to carry.

Pipeline development could be particularly important for shipments of crude oil from Canada to the United States. In light of growing Canadian exports, several proposals have been made to expand the cross-border pipeline infrastructure. Of the five major pipelines currently linking Canadian petroleum producing regions to markets in the United States, two (Alberta Clipper and Keystone) began service in 2010. A permit application for a sixth pipeline, Keystone XL, a very large project which would also transport some Bakken crude, was initially submitted in 2008 and is in the final stages of review by the U.S. Department of State.[107] Keystone XL has been the subject of intense scrutiny and debate by Congress, the executive branch, and numerous stakeholders. The Keystone XL review and approval process is highly contested, and the pipeline's approval remains uncertain.

Other proposed oil pipeline projects, such as the reversal of the Portland-Montreal oil pipeline to enable export of Canadian oil via a marine terminal in Maine, are also encountering greater public scrutiny and opposition. On the whole, the barriers to new oil pipeline approval in any jurisdiction seem to have risen significantly since Alberta Clipper and Keystone were completed.

Shipment of oil by rail is, in many cases, an alternative to new pipeline development. This involves tradeoffs in terms of both transportation capacity and safety. In its ongoing review of the Keystone XL pipeline proposal, the State Department has argued that, if the pipeline is not constructed, additional oil-by-rail capacity will be developed instead. As the State Department's 2014 Final Environmental Impact Statement for the Keystone XL project states,

> In the past 2 years, there has been exponential growth in the use of rail to transport crude oil throughout North America, primarily originating from the Bakken in North Dakota and Montana, but also increasingly utilized in other production areas, including the [Western Canadian Sedimentary Basin]. Because of the flexibility of rail delivery points, once loaded onto trains the crude oil could be delivered to refineries, terminals, and/or port facilities throughout North America, including the Gulf Coast area.[108]

Consistent with this view, both Canadian National Railway and Canadian Pacific Railway reportedly have been pursuing a "pipeline on rails" business strategy, including new track investments, to move Canadian crudes to new markets throughout North America.[109] Increasing cross-border movements of crude oil by rail on existing track does not require State Department approval, so such an approach seeks to avoid regulatory delays. While the potential volumes associated with rail transportation of crude could be lower than pipeline volumes, they could still be significant. Some analysts have suggested that oil-by-rail volumes could be large enough to make a major new pipeline project like Keystone XL unnecessary.[110] Similar arguments could apply to other oil transportation corridors within North America.

Others are less certain that oil by rail can substitute so readily for pipeline capacity, as rail expansion would require significant infrastructure development including loading and unloading facilities, track capacity, and, possibly, additional tank car availability. The State Department's analysis finds that under certain conditions, including particular oil and oil transportation prices, "there could be a substantial impact on oil sands production levels."[111] Other market analyses similarly find that in the short and medium term some production could be curtailed.[112]

Refiner economics may ultimately favor pipelines over rail, although those investment decisions will be determined by market forces. When it comes to safety, however, the federal government plays a major role, and thus may have considerable influence on infrastructure expansion. On balance, it seems likely that policies that raise the cost of transporting oil by rail would increase the attractiveness of pipeline development, and, for that matter, expansion of crude oil transportation by barges, tanker ships, and tanker trucks.

Rail vs. Waterborne Transport

As indicated above, the cost of transporting oil along the coasts in Jones Act tankers is much less than by railroad. However, the fleet of Jones Act-qualified tankers is insufficient to take advantage of this lower-cost shipping method. Despite the domestic oil boom, coastal refineries continue to rely on foreign sources of oil that are shipped at rates generally ranging from less than a dollar per barrel (Mexico to the Gulf Coast) to less than $2.50 a barrel (Saudi Arabia to the Atlantic Coast).[113]

In addition to efficient transport, safety is a concern. Tankers are not a new transport method. Tankers are double-hulled, operators are required to have resources and equipment nearby in case of a spill, and the Coast Guard has a regulatory regime in place to promote safe transits through harbors. While the risk of an oil spill in a marine environment remains a grave concern, coastal transport largely avoids travel through towns and cities. The railroads have had difficulty in dealing with the surge in oil traffic, and other rail users are experiencing severe service disruptions in the upper Midwest. An important but open question is whether more oil would be moving by coastwise shipping, relieving some of the safety and capacity pressure on railroads, if tankers were available and their operating costs more competitive. The CEO of Phillips 66 has stated, "I think our view is that because of the limitations on Jones Act vessels, that's going to push you to more barge and more rail, just to evacuate the Gulf Coast."[114]

Rail Transport and Crude Oil Exports

The large increase in U.S. oil production has led some Members of Congress to advocate changing the law that generally prohibits exports of crude oil.[115] An increase in crude oil exports would likely require greater use of rail transportation, as the crude oil pipeline network is not oriented to serve export ports. Some environmental groups have stated their opposition to construction of new rail facilities or terminals that would facilitate oil exports, as they believe increased exports will encourage environmentally damaging production in the United States and Canada.[116]

End Notes

[1] The terms "oil sands" and "tar sands" are often used interchangeably to describe a particular type of nonconventional oil deposit. Opponents of the resource's development often use the term "tar sands," which arguably carries a negative connotation; proponents typically refer to the material as oil sands. The use of this term is not intended to reflect a point of view, but to adopt the term most commonly used by the primary executive-branch agencies involved in recent oil sands policy issues.

[2] Association of American Railroads, "Moving Crude Oil by Rail," September 2014, p. 4.

[3] Oilgram Price Report, "North American Crude By Rail Rising: BNSF," v. 92, no. 58, March 26, 2014, p. 1.

[4] Estimates by CRS based on data from Dagmar Etkin, Analysis of U.S. Oil Spillage, API Publication 356, August 2009, and Association of Oil Pipelines, Report on Shifts in Petroleum Transportation: 1990-2009, February 2012.

[5] See, for example, Office of Senator John Hoeven, "Hoeven to Meet Saturday with BNSF Railway President and CEO to Address Railroad Safety," press release, January 3, 2014.

[6] Energy Information Administration, U.S. Crude Oil Supply & Disposition, http://www.eia.gov/dnav/pet/ pet_sum_crdsnd_k_a.htm. A barrel of oil is equal to 42 gallons.

[7] CRS Report R43128, Oil Sands and the Oil Spill Liability Trust Fund: The Definition of "Oil" and Related Issues for Congress, by Jonathan L. Ramseur.

[8] "US Crude Production Tops Imports For The First Time Since 1995," Oil Daily, November 14, 2013.

[9] M. Tennyson et al., Assessment of Remaining Recoverable Oil in Selected Major Oil Fields of the Permian Basin, Texas and New Mexico, 2012, USGS, http://pubs.usgs.gov/fs/2012/3051/.

[10] "Light" refers to oils with low specific gravity. "Sweet" refers to oils with low sulfur content. Light, sweet crudes are more valuable than heavier or sourer crude oils.

[11] For further information on the petroleum refining industry, refer to CRS Report R41478, The U.S. Oil Refining Industry: Background in Changing Markets and Fuel Policies, by Anthony Andrews et al.

[12] Energy Information Administration crude oil production data, by state, available at http://www.eia.doe.gov.

[13] For more information about the Keystone XL pipeline, see CRS Report R43787, Keystone XL Pipeline: Overview and Recent Developments, by Paul W. Parfomak et al.

[14] Energy Information Administration price data, available at http://www.eia.doe.gov.

[15] Kirk Eggleston, "Two New Bakken Crude Oil Pipelines Online by 2016, and Enbridge's Sandpiper Pipeline Delayed," Bakken Pipeline and Midstream News, October 31, 2014 and October 2, 2014 available at http://bakkenshale.com.

[16] Pipeline data from PHMSA, railroad mileage from Association of American Railroads (includes shortline rail mileage, does not include parallel trackage).

[17] See 49 C.F.R. §213.9.

[18] STB, Docket no. EP 724, April 10 and September 4, 2014.

[19] BB&T Capital Markets, "Examining The Crude By Barge Opportunity," June 10, 2013, p. 15.

[20] Tank car numbers from presentations by panel on tank car safety (panel 1), NTSB forum, Rail Safety: Transportation of Crude Oil and Ethanol, April 22-23, 2014.

[21] E. Russell Braziel, RBN Energy Inc. presentation at CSIS conference, North American Oil and Gas Infrastructure, Shale Changes Everything," November 14, 2013.

[22] The bow of the tug fits into a notch in the stern of the barge and the tug is hinged to the barge on both sides of its hull, allowing fore and aft (pitch) movement, such as over sea swells.

[23] Platts, Oilgram Price Report, January through November 2014 issues. Railroad distance approximated using Rand McNally Road Atlas.

[24] Platts, Oilgram News, September 9, 2013; Bloomberg Businessweek, "U.S. Law Restricting Foreign Ships Leads to Higher Gas Prices," December 12, 2013. The sailing distance is 1,900 nautical miles (one nautical mile equals 1.151 statute miles); National Oceanic and Atmospheric Administration, Distances Between United States Ports, 2012.

[25] En*Vantage, Inc., "The Surge in U.S. Crude Oil Production," Presentation to PFAA 20th Annual Conference, October 24, 2013; Bloomberg, "Texas Vies with Saudi Arabian Oil in California Shipments," January 29, 2014. When expansion of the locks through the Panama Canal is completed in 2015, the capacity of tankers able to pass through will increase from 380,000 barrels to 600,000 barrels.

[26] Grain and feed producers in the upper Midwest contend that while they can move product economically by barge to New Orleans or by rail to a Great Lakes port, from there, because of the Jones Act, they have no economic access to dry bulk ships that could deliver the feed to eastern North Carolina hog and poultry farms. These farms import their feed from Canada and South America.

[27] The law is codified at chapters 81, 121, and 551 of Title 46, United States Code.

[28] See for instance, "Oil and the Ghost of 1920," Wall Street Journal, September 13, 2012; Senate Committee on Energy and Natural Resources, Testimony of Faisel Khan, Managing Director, Integrated Oil and Gas Research, Citigroup. Hearing to Explore the Effects of

Ongoing Changes in Domestic Oil Production, Refining and Distribution on U.S. Gasoline and Fuel Prices, July 16, 2013.

[29] U.S. Maritime Administration, Title XI Ship Financing Guarantees, Pending and Approved Loan Applications; American Petroleum Tankers S-1 SEC Filing; RS Platou Economic Research, annual and monthly reports; press releases from Teekay Tankers, Scorpio Tankers, and Euronav.

[30] U.S. Maritime Administration, U.S. Flag Privately Owned Merchant Fleet, Oceangoing Self-propelled Vessels.

[31] For further information on the Jones Act as it relates to domestic oil transport, see CRS Report R43653, Shipping U.S. Crude Oil by Water: Vessel Flag Requirements and Safety Issues, by John Frittelli.

[32] "Phillips 66 Charters Tankers To Bring Shale Oil To Bayway," Argus Media, December 13, 2012.

[33] Phillips 66 presentation at Bank of America Merrill Lynch Refining Conference, March 6, 2014.

[34] EIA, Petroleum and Other Liquids, Exports of Crude Oil and Receipts by Tanker and Barge.

[35] This situation is somewhat similar to the Pacific Northwest lumber industry in the 1960s and 1970s, which asserted it could not compete with western Canadian lumber because the Canadians could ship at lower international freight rates to the U.S. East Coast. Washington and Oregon still load significant amounts of wood products on ships, but they all sail to foreign destinations.

[36] Shipping rates from Platts, Oilgram Price Report, January through November 2014 issues.

[37] Sources consulted include NOAA, Oil Spill Case Histories, 1967-1991, Summaries of Significant U.S. and International Spills, 1992; U.S. Coast Guard, Notable Spills in U.S. Waters, Calendar Years 1989-2008, 2009; Dagmar Etkin, Analysis of U.S. Oil Spillage, API Publication 356, August 2009; NOAA, Incident News, at http://incidentnews.gov; EPA, Enforcement and Compliance History Online (ECHO), at http://www.epa-echo.gov/echo/index.html.

[38] National Research Council, Oil in the Sea III: Inputs, Fates, and Effects (Washington, DC: National Academies of Science, February 2003).

[39] These requirements are codified at 49 C.F.R. Section 174.85.

[40] Data from Canada's National Energy Board. See also CRS Report R43128, Oil Sands and the Oil Spill Liability Trust Fund: The Definition of "Oil" and Related Issues for Congress, by Jonathan L. Ramseur.

[41] The primary vehicle for these arguments was a 2011 report from several environmental groups. See Anthony Swift et al., Tar Sands Pipelines Safety Risks, Joint Report by Natural Resources Defense Council, National Wildlife Federation, Pipeline Safety Trust, and Sierra Club, February 2011.

[42] See, e.g., Crude Quality Inc., Report regarding the U.S. Department of State Supplementary Draft Environmental Impact Statement, May 2011; and Energy Resources Conservation Board, Press Release, "ERCB Addresses Statements in Natural Resources Defense Council Pipeline Safety Report," February 2011.

[43] P.L. 112-90, §16.

[44] National Research Council, Effects of Diluted Bitumen on Crude Oil Transmission Pipelines, 2013.

[45] National Transportation Safety Board, Accident Report: Enbridge Incorporated Hazardous Liquid Pipeline Rupture and Release- Marshall, Michigan, July 25, 2010, July 2012, at http://www.ntsb.gov/.

[46] For more up-to-date information, see EPA's Enbridge oil spill website at http://www.epa.gov/enbridgespill/ index.html.

[47] Letter from Cynthia Giles, Environmental Protection Agency, to U.S. Department of State, April 22, 2013.

[48] EPA Removal Order, March 14, 2013, at http://www.epa.gov/enbridgespill/ar/enbridge-AR-1720.pdf.

[49] See Enbridge Inc., Management's Discussion and Analysis, February 2014, at http://enbridge.com/InvestorRelations/FinancialInformation/InvestorDocumentsandFilings.aspx.

[50] Based on cost estimates prepared in 2004. See Dagmar Etkin, Modeling Oil Spill Response and Damages Costs, Proceedings of the 5th Biennial Freshwater Spills Symposium, 2004, at http://www.environmental-research.com.

[51] K.J. Bryden, E.T. Habib, Jr., and O.A. Topete, "Processing shale oils in FCC: Challenges and opportunities," September 1, 2013, http://www.hydrocarbonprocessing.com/Article/3250397/Processing-shale-oils-in-FCCChallenges-and-opportunities.html.

[52] For further information on the properties of Bakken crude oil, see CRS Report R43682, Small Refineries and Oil Field Processors: Opportunities and Challenges, coordinated by Anthony Andrews; and CRS Report R43401, Crude Oil Properties Relevant to Rail Transport Safety: In Brief, by Anthony Andrews.

[53] Pipeline and Hazardous Materials Safety Administration, Safety Alert—January 2, 2014, Preliminary Guidance from OPERATION CLASSIFICATION. This advisory is a follow-up to the PHMSA and Federal Railroad Administration (FRA) joint safety advisory published November 20, 2013 [78 F.R. 69745].

[54] See https://www.dmr.nd.gov/oilgas/OilConditioningOrder25417a.pdf.

[55] http://www.dot.gov/fastlane/state-inspectors-help-fra-improve-rail-safety.

[56] FRA Administrator Szabo, Opening Remarks to RSAC Meeting, October. 31, 2013; http://www.fra.dot.gov/eLib/ Details/L04852.

[57] FRA and PHMSA are agencies within DOT, which has the emergency authority to restrict or prohibit transportation that poses a hazard of death, personal injury, or significant harm to the environment. See 49 U.S.C. §20104.

[58] The NTSB held a forum on the safety of crude oil and ethanol transport by rail April 22 and 23, 2014; http://www.ntsb.gov/news/events/2014/railsafetyforum/.

[59] 40 C.F.R. Part 112.

[60] 40 C.F.R. §300.5. This definition is also found in the Clean Water Act and OPA.

[61] The Oil Pollution Act of 1990 expanded and clarified the President's authorities under Section 311 of the Clean Water Act (33 U.S.C. §2701 et. seq.). For a more in-depth discussion of the Oil Pollution Act, see CRS Report RL33705, Oil Spills in U.S. Coastal Waters: Background and Governance, by Jonathan L. Ramseur.

[62] 33 U.S.C. §1321. In addition, the Comprehensive Environmental Response, Compensation, and Liability Act (CERCLA) of 1980 expanded the authorities of the President to respond to releases of hazardous substances into the environment more broadly than CWA Section 311. See CRS Report R41039, Comprehensive Environmental Response, Compensation, and Liability Act: A Summary of Superfund Cleanup Authorities and Related Provisions of the Act, by David M. Bearden. For further details, see CRS Report R43251, Oil and Chemical Spills: Federal Emergency Response Framework, by David M. Bearden and Jonathan L. Ramseur.

[63] 33 U.S.C. §1321(c).

[64] Executive Order 12777, "Implementation of Section 311 of the Federal Water Pollution Control Act of October 18, 1972, as amended, and the Oil Pollution Act of 1990," 56 Federal Register 54757, October 22, 1991.

[65] 49 C.F.R. Part 130.

[66] See 49 C.F.R. §130.31(a) with 49 C.F.R. §130.31(b).

[67] 61 Federal Register 30541 (June 17, 1996).

[68] NTSB, Safety Recommendation R-14-4 through -6, directed to PHMSA, January 21, 2014.

[69] 79 Federal Register 45079 (August 1, 2014).

[70] See FRA's Emergency Order No. 28 (78 Federal Register 48218), the agencies' Joint Safety Advisory published August 7, 2013 (78 Federal Register 48224), referral of safety issues to FRA's Railroad Safety Advisory Committee (78 Federal Register 48931), and a NPRM related to rail hazmat (78 Federal Register 54849).

[71] Xiang Liu, M. Rapik Saat, Christopher P.L. Barkan, Safety Effectiveness of Integrated Risk Reduction Strategies for the Transportation of Hazardous Materials by Rail, paper presented at the Transportation Research Board, Annual Meeting 2013, paper no. 13-1811.

[72] AAR, "Freight Railroads Join U.S. Transportation Secretary Foxx in Announcing Industry Crude By Rail Safety Initiative," February 21, 2014.

[73] 79 Federal Register 45016-45079 (August 1, 2014); http://www.regulations.gov, docket no. PHMSA-2012-0082. The comment period closed September 30, 2014; 3,304 separate comments were filed.

[74] The tank cars used to transport crude oil fall under DOT specification 111. See 49 C.F.R. §179.

[75] 49 C.F.R. §179.201.

[76] For a discussion of NTSB's recommendations concerning DOT-111 tank cars, in reference to the derailment of an ethanol unit train in Cherry Valley, IL, see NTSB Safety Recommendation R-12-5 through -8, March 2, 2012.

[77] Pressurized tank cars (DOT specification 105 and 112) have thicker shells and heads, metal jackets, strong protective housings for top fittings, and no bottom valves.

[78] 78 F.R. 54849-54861, September 6, 2013.

[79] For comments filed on this rulemaking see http://www.regulations.gov and search under docket no. PHMSA-2012- 0082.

[80] See letter from Senator Schumer to PHMSA and FRA dated July 22, 2013 and news release by Senator Hoeven on January 15, 2013 indicating that a DOT final rule on tank cars would not be issued until after January 2015.

[81] Transport Canada, "Addressing the safety of DOT-111 tank cars carrying dangerous goods," April 23, 2014.

[82] 79 Federal Register 45016-45079 (August 1, 2014).

[83] See comments filed by the American Petroleum Institute, September 30, 2014; http://www.regulations.gov, docket no. PHMSA-2012-0082.

[84] See comments filed by the Association of American Railroads, September 30, 2014; http://www.regulations.gov, docket no. PHMSA-2012-0082.

[85] T87.6 Task Force Summary Report, pp. 9-11; Xiang Liu, M. Rapik Saat, Christopher P.L. Barkan, "Analysis of Causes of Major Train Derailment and Their Effect on Accident Rates," Transportation Research Record, No. 2289, 2012, pp. 154-163.

[86] Xiang Liu, M. Rapik Saat, Christopher P.L. Barkan, Safety Effectiveness of Integrated Risk Reduction Strategies for the Transportation of Hazardous Materials by Rail, Paper presented at the Transportation Research Board, Annual Meeting 2013, paper no. 13-1811.

[87] 79 Federal Register 4234, January 24, 2014.

[88] Xiang Liu, Broken "Rail Prevention and Rail Flaw Assessment," presentation at Transportation Research Board Annual Meeting, Washington, DC, January 13, 2014.

[89] 79 Federal Register 4234 and 4245, January 24, 2014.

[90] Presentations of BNSF and UP Railroads, Transportation Research Board Annual Meeting, Washington, DC, January 13, 2014, Session 279, Broken Rail Prevention and Rail Flaw Assessment.

[91] Christopher P.L. Barkan, M. Rapik Saat, and Francisco Gonzalez III et al., "Cooperative Research in Tank Car Safety Design," TR News, vol. 286 (May-June 2013), pp. 12-19.

[92] This topic was briefly discussed at the NTSB forum on rail safety cited above. Advances in detecting internal rail flaws are focused on increasing testing speeds, shrinking the size of detectable flaws, and improving the training of testing operators; "Tracking Down Rail Flaws," Railway Track and Structures, January 2011, pp. 19-22.

[93] The Railroad Rehabilitation and Improvement Financing (RRIF) program and Section 45G of the tax code.

[94] Athaphon Kawprasert and Christopher P.L. Barkan, "Effect of Train Speed on Risk Analysis of Transporting Hazardous Materials by Rail," Transportation Research Record, No. 2159, 2010, pp. 59-68.

[95] 79 Federal Register 45016-45079 (August 1, 2014).

[96] See comments filed by the Association of American Railroads, September 30, 2014; http://www.regulations.gov, docket no. PHMSA-2012-0082.

[97] In a separate NPRM (79 Federal Register 53356-53383, September 9, 2014), U.S. DOT addressed brake application systems for securing unattended parked trains, which was the issue in the Lac Mégantic, Quebec, incident (see text box).

[98] See comments filed by the Association of American Railroads, September 30, 2014; http://www.regulations.gov, docket no. PHMSA-2012-0082. ECP brakes were proposed by the FRA in a 2008 rulemaking; 73 Federal Register 61512 (October 16, 2008).

[99] Press Release no. FRA-03-14, April 9, 2014.

[100] NTSB, Collision of Two CN Freight Trains, Anding, Mississippi, July 10, 2005, Accident Report RAR-07/01, p. 31.

[101] For further information, see CRS Report R42637, Positive Train Control (PTC): Overview and Policy Issues, by John Frittelli.

[102] See 49 C.F.R. §172.820; 73 Federal Register 72182, November 26, 2008.

[103] See 49 C.F.R. §1580.107.

[104] U.S. Rail News, June 11, 2008, pp. 1-2; "Hazmat Hazards: U.S. Cities may not wait for Washington Before Trying to Reroute their own hazmat trains," Journal of Commerce, December 12, 2005.

[105] 79 Federal Register 45042, August 1, 2014. The NTSB has recommended this change; see Safety Recommendation R-14-1 through -3, January 23, 2014.

[106] Section 33005 of P.L. 112-141.

[107] The construction, connection, operation, and maintenance of a pipeline connecting the United States with a foreign country require executive permission through a Presidential Permit under Executive Orders 11423 and 13337.

[108] U.S. Department of State, January 2014, Final EIS, Section 5.1, "No Action Alternatives."

[109] Nathan Vanderklippe, "CN, CP Push for a 'Pipeline on Rails,'" Globe and Mail, February 7, 2011.

[110] "Keystone Pipeline Seen as Unneeded as More Oil Moves by Rail," CBC News, September 10, 2013.

[111] 2014 Final EIS, p. 1.4-8.

[112] For example, Canadian Imperial Bank of Commerce, “Too Much of A Good Thing: A Deep Dive Into The North American Energy Renaissance,” August 15, 2012; TD Economics, “Pipeline Expansion is a National Priority,” Special Report, December 17, 2012; International Energy Agency," Medium-Term Oil Market Report," May 14, 2013.

[113] Platts, Oilgram Price Report, January through November 2014.

[114] Phillips 66 Earnings Conference Call, Q2 2013, July 31, 2013.

[115] The Senate Energy and Natural Resources Committee held a hearing on this issue on January 30, 2014. For further information and analysis on oil exports, see CRS Report R43442, U.S. Crude Oil Export Policy: Background and Considerations, by Phillip Brown et al.

[116] See, for example, the comments of Sierra Club official Michael Marx in Blake Sobczak, “Environmentalists ‘get real creative’ to combat oil by rail,” Energy Wire, January 13, 2014.

INDEX

D

E

F

G

H

I

J

K

L

Q

R

U

V

W

Y